Ecological Integrity in Science and Law

Laura Westra • Klaus Bosselmann
Matteo Fermeglia
Editors

Ecological Integrity in Science and Law

Springer

Editors
Laura Westra
University of Windsor
Maple, ON, Canada

Klaus Bosselmann
Faculty of Law
University of Auckland
Auckland, New Zealand

Matteo Fermeglia
Faculty of Law
Hasselt University
Hasselt, Belgium

ISBN 978-3-030-46258-1 ISBN 978-3-030-46259-8 (eBook)
https://doi.org/10.1007/978-3-030-46259-8

This Springer imprint is published by the registered company Springer Nature Switzerland AG.
The registered company address is: Gewerbestrasse 11, 6330 Cham, Switzerland

Preface

This year we have decided to return to our roots as the Global Ecological Integrity Group which started as a small group of scientists and one ethicist before evolving into an interdisciplinary network. At our meeting at the University of Salerno in 2018, science scholars among us suggested to focus on methodologies for the next meeting at the University of Trieste, Italy, in 2019. We were successful to attract many scientists attending the meeting and contributing to this book.

The present collection includes many from different disciplines who enjoy a worldwide reputation. The first such scholar is Donald Brown, whose work on climate change has netted him the Avicenna Prize, just awarded to him. His chapter analyzes the IPCC work on climate change and finds it severely lacking in ethical analysis and even the acknowledgment of existing soft law. The author of the second chapter Virginia Zambrano reprises the theme of climate change from the standpoint of industrial/animal agriculture, not only for its role in climate change through the presence of methane that results but also because of the hazardous chemical inputs that industry depends on.

Matteo Fermeglia addresses climate change in a more positive vein. He discusses three cases where climate change litigation acknowledged to some extent the responsibility of the respective states to change their policies. The very fact of accepting some responsibility for the present impasse on the part of state administrations or courts gives hope for further accountability on the part of policy-makers.

Janice Gray examines a specific issue in her country. She establishes the presence of the Water Act 2007 and the presence of specific plans addressing the Murray–Darling Basin, which should have protected the area's waters. Yet, she says, the water "has been stolen by some irrigators, traded commercially" and the local fish have been exterminated. In addition, "water buybacks" have taken place in ways the author defines as "shady deals." Gray concludes that even the presence of science-based law should be closely monitored in order to ensure compliance.

The final chapter in Part I by Joseph Dellapenna deals with a grave challenge for what he calls the "frontline states" in the Mediterranean Sea, as thousands of migrants/refugees face not only a perilous journey but also a high possibility of death. The line between grants and refugees is not easy to draw in most of these cases. There have been migrations in all parts of the world in the twentieth century, but the UN Refugee system seems to have been unable to cope with the present situation. Despite the presence of regulatory regimes such as Dublin III, Regulation 2013, the sheer numbers have led to a collapse of the EU migration system. In some cases like that of Italy's Matteo Salvini, leader of the Lega, he simply decided to close all parts and give in to populist demands for "sealing the borders." Dellapenna concludes that the protection of refugees' rights is at great risk at this time.

Part II addresses the issue of public health. Carl Cranor discusses the presence of numerous unchecked chemicals everywhere, posing a grave risk to the biological integrity of persons. Cranor differentiates between infectious diseases and chronic ones, and the fact that the latter can sometimes be treated but not always cured. Patients are recommended to avoid risky behaviors. The main issue is that chronic diseases are often caused by products or actions of others, and justice requires that the risk to others should not be permitted. There are some laws intended to prevent these exposures. But not enough to make a difference, as the scientific evidence demonstrates that, as far as chronic diseases are concerned, there are clear "developmental origin of disease" and re-birth exposures mayhem not only children but adults as well. Current legal regimes are insufficient and science should play a stronger role in dictating appropriate policies.

The next chapter is by Colin Soskolne who chaired the session on Public Health. Soskolne, a career occupational and environmental cancer research methodologist, addresses the obligation of health scientists to protect public health and safety from established, as well as suspected, avoidable harms. The prevention of illness and premature death is central to public health's mission. Being a soft science, epidemiology provides the core methods for public health protection. Because epidemiological evidence can be influenced by interests that manipulate the scientific method to cast doubt, thus fomenting uncertainty designed to mislead both the public and policy-makers, Soskolne describes some ethical and practical dimensions of influence that result in delayed policy action. He provides glimpses of some instances with ways of addressing the challenges posed. To protect the public interest, he emphasizes the need for vigilance by professionals with the capacity to speak truth to power; he notes the need to train students to diagnose manipulation and provide strategies for calling deception to account.

The final chapter in Part II has Shauna Lange emphasizing the role of women in the achievement and protection of ecologically valuable goals. From Jane Goodall's gorillas to Greta Thunberg's amazing work rallying children and young adults to environmental causes, many women have fought to denounce environmental crimes, and some have lost their life doing so. Lange proceeds to discuss what precisely is shared by the women she names. She suggests perhaps the tendency to be global thinkers, to fight traditional cultural roles, to tend to be caregivers, this being closer to nature than men are. Both states and local governments are most often run primarily by men, who often do not share the ultimate goal to view humanity as a whole as a species worth preserving, as women end to do.

Part III considers possible obligations on the part of states to protect ecological integrity. In the first chapter of this part, Ngozi Unuigbe presents the connection between African eco-philosophy and ecological integrity. Africa, she notes, has numerous grave environmental and human rights problems. The local version of eco-philosophy, eco-relationism, is handed down as a traditional religion of Africa through the generations: it is based on the belief that current life, both human and non-human, present, past, and future are all interconnected. Hence our overconsumption places at risk the "biotic community," she states, citing Aldo Leopold. Thus, clearly, individual interests cannot come before the rights of the community of nature.

Kathryn Gwiazdon addresses the responsibility of the state regarding the ongoing environmental crisis. She cites Einstein's strong critique of war and his demand that human rights should be put ahead of all, even state sovereignty rights. She points out the lack of ethical standards, social bods, and the very breakdown of social order as part of the reasons that render so difficult the acceptance of ethical limits for all, while worldwide the failures of international law to offer redress persist. She argues that the denial of reality leads to accept and embrace injustice. Gwiazdon concludes that state sovereignty today and the present global governance are simply not sustainable.

Geoffrey Garver's chapter proposes returning to individual and community's attachment to places, to counteract the present separation of humans from nature. What is needed, Garver argues, is increased respect for the Indigenous Peoples whose attachment to their places is part of their tradition. Humanity as a whole needs to connect to place, to substitute the uncaring mobility and detachment that prevails.

The next chapter asks whether it is still possible to protect wilderness today. Anne Venton notes that despite the fact that parks were protected in many parts of the world, today the situation is not so positive. She cites some who see the impending approach of an "environmental rights revolution," which would include the constitutional right to a healthy environment. The growing public awareness of climate change recently resent in most parts of the world is a case in point, which permits us to hold a moderately optimistic outlook.

The last chapter in Part III considers what we can learn from science, specifically quantum mechanics. Orsolya Banyai argues that the resent laws are "not consistent with the fundamental law and principles" that direct the functioning of the universe. Too many of the planetary boundaries have been transgressed: climate change, losses of biodiversity are clear indications. But she cites the existence of the Great Jurisprudence, which transcends even the approach proposed by natural law, while both in fact also correspond to some Eastern classic theories, regarding the universe. These "higher rules" beyond human laws are what is needed to correct present legal regimes.

Part IV returns to the role of earth science. Jakub Zelený proposes a land-based approach to ecological integrity, which represents a wonderful addition to the way we have been understanding both biological (single organism) and ecological (general) integrity. In Canada, we started from the 1972 Clean Water Act, so that water actually permitted both considerations, when adding the biota that lived with it. Zelený proposes both a new methodology and additional considerations, starting from the scientific research based on land instead.

In the next chapter, Eva Cudlínová discusses how wilderness can be protected in the Czech Republic, taking into consideration that even the United Nations cited the protection of Nature as one of its Sustainable Development Goals. She discusses the situation in Sumava National Park, reflects on the understanding of wilderness in the USA and the EU, and finds that in most cases the definition of wilderness zones involves an area with "effective functioning of ecological Processes…It is unmodified or only slightly modified…without intrusive or extractive human activity" (Wild Europe 2019). Cudlínová cites also the main conflicts that arise in the very presence of such parks, including zoning and cultural opposition to their presence.

Pavel Cudlin, in the next chapter, even raises the difficult question of whether it is even possible to maintain the biodiversity required for the establishment of wilderness areas under conditions of increasing landscape exploitation. In the Czech Republic, there have been many areas of meadows and pasture, but much smaller areas devoted to forests, as natural and near-natural habitats, but all are in decline due to landscape exploitation and climate change. His chapter then analyzes land use and biodiversity changes and concludes that the persistence of natural and near-natural habitat is now quite low.

In the final chapter in this section, Alicia Epstein discusses the European Union Common Agricultural Policy (CAP), which she defines as governed by years of economic austerity. The continuous yields in agriculture in the EU have been enabled "by the employment of highly intensive management and production practices." But the intensive use of pesticides and fertilizers has contributed to the disruption of the Earth's phosphorus and nitrogen cycles, and these intensive practices were employed without any consideration of long-term environmental effects. Although plans of the CAP are said to reflect "strong sustainability principles," it will necessitate a complete transformation of the present practices of CAP.

In Part V, Peter Venton discusses the necessity of managing the economy without growth. He considers the political economy of the Canadian government, as he argues that both climate change and poverty itself would be significantly reduced if a policy model based on business as usual were abandoned. He examines the various theories that influence voter participation in a democracy, but adds that present democratic institutions have abandoned the quest for the common good, whereas present democracy bears an uncomfortable similarity to oligarchy instead, since economic growth seems to precede all other goals, whereas there would be significant gains for all if a no-growth system of governance was adopted.

Prue Taylor's chapter is focused on her recent work on the "responsibility to protect" (R2P) doctrine. R2P emerged in response to the humanitarian disaster in Rwanda and has been described as a tectonic shift in state sovereignty. Arguably, states have a legal obligation to protect people within their jurisdiction and if failing to do so, the international community has a responsibility to assist. The same rationale, Taylor argues, applies to the legal obligation of states to protect people from environmental harm. R2P should therefore inform any initiatives to foster environmental responsibilities of states such as the current efforts towards a Global Pact for the Environment.

The underpinning governance aspects of Taylor's chapter are the subject of Klaus Bosselmann's chapter on environmental trusteeship. The system of international environmental law is still predicated by a concept of state sovereignty that is ecologically blind and unfit for purpose, i.e. protecting people and planet from harm. Yet, neither state sovereignty nor international law are static constructs. As GEIG's work over the years has shown, ecological integrity is not just sound science with respect to measuring ecosystem health, many international environmental agreements express the preservation of Earth's ecological systems a duty of states. It is now possible to describe Earth trusteeship as a key function state governance as, for example, articulated in the "Hague Principles" (2018).

Gabriella Colao explores the rules and criteria developed and implemented to ensure remediation of environmental damages in times of warfare. The lack of an integrated and holistic approach in the normative framework applicable to armed conflict renders the environment still extremely vulnerable and at risk. Hence, Colao firmly advocates the need to shift towards an inherently ecological approach to environmental protection during war conflicts while embracing the ecosystem as a common good, as well as a revisited notion of national interest and responsible sovereignty.

Toronto, ON
January 4, 2020

Laura Westra
Klaus Bosselmann

Contents

Part I Ecological Integrity: Climate Change and Water Issues

Lessons Learned from IPCC's Underestimation of Climate Change Impacts About the Need for a Precautionary Climate Change Science 3
Donald A. Brown

Animal Based Industries and Climate Change 11
Virginia Zambrano

Climate Science Before the Courts: Turning the Tide in Climate Change Litigation 23
Matteo Fermeglia

'Thieves, Shady Deals and Murder': Water Theft, Buy-Backs and Fish Kills in the Murray Darling Basin of Australia 37
Janice Gray

Refugees at Sea in a Hostile World 51
Joseph W. Dellapenna

Part II Public Health and Ecointegrity

Using Science and Law to Promote Humans' Biological Integrity 65
Carl F. Cranor

Epidemiology and Public Health Under Siege: In Whose Best Interests? 77
Colin L. Soskolne

Saving Species, Healthy Humanity: The Key Role of Women in Ecological Integrity 85
Shauna M. Lange

Part III Ecological Integrity and State Obligations

African Eco-Philosophy and Its Implications for Ecological Integrity in Africa 99
Ngozi Finette Unuigbe

State Global Responsibility for Environmental Crises: The Ethical and Legal Implications of a State's Failure to Protect Human Rights 111
Kathryn Anne Gwiazdon

Promoting Human Attachment to Place in Ecological Law 125
Geoffrey Garver

Protected Wilderness: Is It Still Possible? 135
Anne Venton

Quantum Mechanics and Law: What Does Quantum Mechanics Teach Us? 147
Orsolya Bányai

Part IV Ecological Integrity and the Role of Earth Science

Evaluation of Ecological Integrity in Landscape Based on Remote Sensing Data 161
Jakub Zelený and Daniel Mercado-Bettín

Social and Economic Conditions for Wilderness Protection in Europe: Case of Šumava National Park, Czech Republic 175
Eva Cudlínová, Miloslav Lapka, and Martin Šlachta

Is it Possible to Maintain a Biodiversity of Natural Habitats Under Global Change and Increasing Landscape Exploitation? 185
Pavel Cudlín, Vilém Pechanec, Marcela Prokopová, Lenka Štěrbová, Jan Purkyt, and Ondřej Cudlín

Agriculture in the European Union: Seven More Years of Environmental Austerity? 201
Alicia A. Epstein

Part V Ecological Integrity and the Geopolitical Situation

The Political Economy of Managing Without Growth 213
Peter Venton

Opportunity Within Failure: Can the Global Pact for the Environment Learn from Responsibility to Protect? 227
Prue Taylor

The Role of Trusteeship in Earth Governance 241
Klaus Bosselmann

Environmental Damages and Armed Conflict 253
Gabriella Colao

Part I
Ecological Integrity: Climate Change and Water Issues

Lessons Learned from IPCC's Underestimation of Climate Change Impacts About the Need for a Precautionary Climate Change Science

Donald A. Brown

Abstract Several recent studies reveal that the International Panel on Climate Change (IPCC), the organization created by the international community to synthesize the peer-reviewed climate change science to provide governments with objective, scientific understanding of climate change, its natural, political and economic impacts and risks, and possible response options has been underestimating likely climate change impacts. Analysis of the causes of IPCC's failure to identify the likely worst climate change impacts attributes the IPCCs underestimation of climate change impacts to: (a) the "consensus" methods of IPCC processes and, (b) that the relevant sciences follow epistemic norms designed to prevent false positive conclusions about cause and effect. This chapter argues that given the enormous potential harms from climate change, a precautionary science that allows scientists to identify all scientifically plausible harms is required by ethics and international legal principles.

1 Recent Evidence That IPCC Is Underestimating Climate Change Impacts

The Intergovernmental Panel on Climate Change (IPCC) is the UN body established for assessing the sciences related to climate change. It was established by the United Nations Environment Programme and the World Meteorological Organization in 1988 to provide policymakers with regular scientific assessments of climate change issues, its implications and potential future risks, as well as to put forward adaptation and mitigation strategies. It has 195 member states (IPCC 2019).

D. A. Brown (✉)
Widener University, Commonwealth Law School, Harrisburg, PA, USA

L. Westra et al. (eds.), *Ecological Integrity in Science and Law*,
https://doi.org/10.1007/978-3-030-46259-8_1

IPCC assessments provide governments, at all levels, with scientific information that they can use to develop climate policies. IPCC assessments are a key input into the international negotiations to tackle climate change. IPCC reports are drafted and reviewed in several stages, thus guaranteeing objectivity and transparency (IPCC 2019). Because of the transparency of its assessment processes and the quality of its scientific assessment, IPCC's analyses are widely respected around the world and among scientific organizations whose members engage in research relevant to climate change policy issues.

The IPCC assesses the thousands of scientific papers published each year to tell policymakers what we know and don't know about the risks related to climate change. The IPCC identifies where there is agreement in the scientific community, where there are differences of opinion, and where further research is needed. It does not conduct its own research (IPCC 2019).

In October of 2018, IPCC issued a Special Report on limiting warming to the Paris Agreement's warming limit goals of as close as possible to 1.5 °C but no more than 2.0 °C (IPCC 2018).

This report concluded that significant climate impacts already occur at 1.5 °C, especially in regard to low-lying areas, human health, and oceans (WRI 2018). These impacts will hit the poor and most vulnerable the hardest due to loss of livelihoods, food insecurity, population displacement, health effects and other causes of social disruption. This report also compared the climate change harms from 1.5 °C warming compared to harms caused by 2.0 °C warming.

This Special Report of IPCC on 1.5 °C Warming received attention around the world for alerting many that climate change was more threatening than expected. See for instance, the "National Geographic Report, "Climate change impacts worse than expected" which explained how the IPCC Special Report made many aware that climate change was more ominous than they previously understood (National Geographic 2018).

Since the IPCC Special Report was published, several scientific reports have been published which support the conclusions that the IPCC conclusions have been underestimating likely climate change impacts.

On July 31, 2018, a paper was published in the Proceedings of the National Academy of Sciences which should create a shiver of fear in all humans everywhere. The paper, "Trajectories of the Earth System in the Anthropocene" by Steffen et al. explains how human-induced warming is rapidly approaching levels that may trigger positive climate feedbacks which could greatly accelerate the warming already plaguing the world by causing record floods, deadly heat waves and droughts, increasing tropical diseases, forest fires, more intense and damaging storms, sea level rise, coral bleaching, and acidification of oceans, all of which are contributing to increasing the number of refugees which are destabilizing governments around the world.

Another paper published in August 2018 concluded that models used to predict climate impacts have failed to incorporate abrupt carbon feedbacks from permafrost decay that recent evidence has revealed are now possible (Anthony et al. 2018). In fact, this paper claims that early stages of processes that lead to permafrost degradation are already underway, a phenomenon which could lead to release of dangerous amounts of methane and CO_2. This paper also concluded that carbon emissions from melting permafrost could increase soil carbon emissions by 125–190% compared to gradual thaw alone. This paper also concluded that models used to predict climate impacts have failed to incorporate abrupt carbon feedbacks from permafrost decay that recent evidence has revealed is now possible. Carbon emissions from melting permafrost could increase soil carbon emissions by 125–190% compared to gradual thaw alone.

Another paper published on May 20, 2019 in the Proceedings of the National Academy of Sciences, "Ice sheet contributions to future sea-level rise from structured expert judgment" (Bamber et al. 2019), concluded that sea level rise could be 2 m by 2100, double the amount predicted by IPCC in its Fifth Assessment Report.

Another paper published in February 2019 in the Journal of Biogeochemical Cycles, "Very Strong Atmospheric Methane Growth in the 4 Years 2014–2017," (Nesbit et al. 2019) warned of greater than expected atmospheric methane levels. This paper claimed that the rise in atmospheric methane (CH4), which began in 2007, had accelerated in the past 4 years. This growth has been worldwide, especially in the tropics and northern midlatitudes. With the rise has come a shift in the carbon isotope ratio of the methane. According to this paper, methane's increase since 2007 was not expected in future greenhouse gas scenarios compliant with the targets of the Paris Agreement, and if the increase continues at the same rates it may become very difficult to meet the Paris goals.

2 Why Is IPCC Underestimating Climate Impacts?

Another paper, "What Lies Beneath: On the Understatement of Existential Climate Risk" (hereinafter "WLB" 2018) by D. Sprat & I. Dunlap and published by the Breakthrough Institute, claims both that the risks posed by climate change are far greater than is evident from the conclusions of IPCC and examines why IPCC has frequently underestimated threats from climate change.

The WLB report further concludes that climate change is now an existential risk to humanity, that is an adverse outcome that could either annihilate intelligent life or permanently and dramatically curtail its potential (WLB 2018, p. 13).

Although the WLB report acknowledges IPCC has done "critical, indispensable work of the highest standard in pulling together a periodic consensus of what must be the most exhaustive scientific investigation in world history" however, WLB claims that the IPCC process suffers from all of the dangers of consensus-building in such a wide-ranging and complex arena (WLB 2018, p. 5). That is, the report attributes the overly conservative conclusions of the IPCC to the consensus nature that IPCC must follow to get governments to approve IPCC final reports and the pressure that some governments would place on IPCC if some conclusions were not based on scientifically rigorous evidence.

The WLB report also claims that, quoting a 2014 article in the Guardian, increasing evidence 'that policy summaries on climate impacts and mitigation by the IPCC were significantly "diluted under political pressure from some of the world's biggest greenhouse gas emitters, including Saudi Arabia, China, Brazil, and the United States." (WLB, p. 34).

This WLB report also attributes IPCC's underestimation of climate impacts to IPCC's synthesizing peer-reviewed science that is based on scientific norms that condemn speculation (WLB 2018, p. 5). For example, although scientists knew that permafrost which contains methane could leak melt from the top down and the bottom up, because scientists could not observe how the permafrost would melt from the bottom up they could not develop empirically verifiable risk assessments of permafrost melting from the bottom up. As a result, the report concludes that some of the climate research on which IPCC has relied has tended to underplay climate risks and as a result, IPCC has exhibited preferences for conservative estimates of climate change impacts (WLB 2018, p. 5). This practice the WLB reports labels as "scholarly reticence." (WLB 2018, p. 5).

This WLB report further claims that climate science has succumbed to the norm followed by most physical sciences to refrain from any speculation that cannot be grounded in empirically determined probability calculations. This epistemic norm, the report claims, is not well-suited to guide predictions about very scientifically complex matters such as earth system dynamics. The report calls this approach the *Probability Obsession* of science which is not well suited to predict future states of complex systems about matters for which there are no historical antecedents (WLB 2018, p. 2).

The WLB report also notes that a conservative approach to climate science began to dominate and as a result, the planetary future has become a hostage to national economic self-interest. Thus, the paper claims it became "alarmist" to claim the climate change is an existential threat to life on earth (WLB 2018, p. 4).

The report further notes that although "a fast emergency-scale transition to a post-fossil fuel world is absolutely necessary to address climate change.... yet this is excluded from consideration by policymakers because it is considered to be too disruptive." And so the paper claims "we have a policy failure of epic proportions." (WLB 2018, p. 4).

The WLB report further notes that although it has widely been reported that if the ghg emissions reductions commitments or Nationally Determined Commitments (NDCs) made by governments so far under the Paris Agreement are complied with, the Earth's temperature is expected to rise to 3.4 °C by 2100 without taking into account "long-term" carbon cycle feedbacks (WLB 2018, p. 15). Yet if the positive feedbacks are fully considered, the temperature path defined by the NDCs could result in around 5 °C of warming by 2100 according to a MIT study (WLB 2018, p. 13). Yet, the report claims that even if warming reaches 3 °C, most of Bangladesh and Florida would drown, while major coastal cities—Shanghai, Legos, Mumbai—would be swamped likely creating larger flows of climate refugees. Most regions of the world would see a significant drop in food production and an increasing number of extreme weather events, whether heat waves, floods or storms (WLB 2018, p. 13).

The WLB also report concludes warming of 4°C or more could reduce the global human population by 80% or 90%, and the World Bank reports "there is no certainty that adaptation to a 4 °C temperature rise would be possible." Quoting Professor Kevin Anderson, the report claims a 4 °C future "is incompatible with an organized global community and is likely to be beyond adaptation by the majority of people." (WLB 2018, p. 14).

The WLB report also claims that the often-quoted prediction of likely temperature increases if current NDCs are complied with of approximately 3 °C rise does not take into account the considerable risk that self-reinforcing feedback loops could be triggered when certain thresholds are reached leading to an ever-increasing rise in temperature. These potential thresholds include the melting of the Arctic permafrost releasing methane into the atmosphere, forest dieback releasing carbon currently stored in the Amazon and boreal forests, with the melting of polar ice caps that would no longer reflect the light and heat from the sun (WLB 2018, p. 14).

The report cites a recent study by the European Commission's Joint Research Center found that if global temperature rose to 4 °C that extreme heat waves with "apparent temperatures" peeking over 55 °C (131 °F) will begin to regularly affect many densely populated parts of the world, forcing much activity in the modern industrial world to stop (WLB 2018, p. 14). The paper claims that one study found that even a 2 °C warming "would double the land area subject to deadly heat and expose 48% of the population to deadly heat." (WLB 2018, p. 14).

According to the WLB report, a 4 °C warming by 2100 would subject 47% of the land area and almost 74% of the world population to deadly heat which could pose existential risks to humans and mammals alike unless massive adaptation measures are implemented (WLB 2018, p. 14).

The WLB paper also explains how IPCC's understatements of likely climate change impacts affect what is generally claimed among climate policy-makers about elements of climate science including climate models, climate tipping points, climate sensitivity, carbon budgets, permafrost and carbon cycles, arctic sea ice, polar ice-mass loss, and sea-level rise.

3 The Ethical Duty to Engage in Precautionary Climate Change Science

Opponents of climate change policies have successfully convinced governments and many citizens around that government action to prevent potential harms not fully proven is unjustified because such actions are based on "bad science."

Ethics would require higher levels of proof of those who are engaged in dangerous behavior to prove their behavior is safe in proportion to how potentially dangerous the behavior is especially for harms to others who have not consented to be harmed and for behaviors that become more dangerous the longer one waits to reduce the uncertainty. International environmental law also contains principles that failing to take action to protect people from activities that are known to be dangerous on the basis of scientific uncertainty is not a legally acceptable justification for preventing the harm. In fact the 1992 United Nations Framework Convention on Climate Change contains the precautionary principle which states:

> The Parties should take precautionary measures to anticipate, prevent or minimize the causes of climate change and mitigate its adverse effects. Where there are threats of serious or reversible damage, lack of full scientific certainty should not be used as a reason for postponing such measures, taking into account that policies and measures to deal with climate change should be cost-effective so as to ensure global benefits at the lowest possible cost. (UNFCCC 1992, Art. 3.3)

The duty to be careful to not harm others even in cases where the proof of harm is uncertain is widely accepted around the world by such international law principles as the "No Harm Principle." Under the No Harm Principle, nations are expected to prevent potential harm to other nations once they have reason to believe that activities in their countries are putting others at risk, they may not wait until absolute proof has been established to cease dangerous behavior (Brown 2019).

Given that climate change actually threatens life on Earth including billions of people who have not consented to put at risk, and given that waiting to reduce ghg emissions makes the problem more threatening, ethics would shift the burden of proof to those who are most responsible for raising ghg emissions to prove with very high levels of proof that human emissions of ghg are safe even if there is some uncertainty about the amount of warming that different levels of ghg emissions will cause. For this reason, the problem created by IPCC's underestimation of climate change impacts may not be exclusively the fault of IPCC. The problem may also be the fault of policymakers who fail to respond to the enormous potential harms entailed by human-induced warming by demanding that opponents of climate change policies shoulder the burden of proof by demonstrating with high levels of proof that ghg emissions will not cause serious harms.

Yet, policymakers have a vital need for scientists to explain all scientifically plausible harms that may result from human activities even if the magnitude and creation of potential harms are uncertain. In fulfilling these responsibilities, scientists may not ignore potential harms because they are unable to determine probabilities about the likelihood of their occurrence based on empirical observations. Yet because scientists often follow the epistemic norms of their science when engaged in scientific research which norms usually require adequate levels of proof before making causal claims, policymakers need to be clear when interacting with scientists that their policymaking responsibilities require that they, the policymakers, protect citizens from all plausible harms. Therefore policymakers need scientists to identify all scientifically plausible harms.

Because IPCC's mission is to synthesize the existing peer-reviewed climate science, which usually does not include scientific conclusions about plausible harms derived from speculation, IPCC cannot fulfill the role of science that policymakers need when policymakers are seeking to protect citizens from all plausible harms and to inform humanity about all plausible climate change impacts. Thus, there is a basic conflict between IPCC's mission of synthesizing peer-reviewed climate change science and providing policy-makers with information about all scientifically plausible climate change impacts.

This need of policy-makers to understand all plausible harms creates an enormous challenge for mainstream scientific institutions which usually rely on peer-review in which scientists normally review scientific claims by comparing claims to empirically tested observations *which are the ground of the scientific enterprise.* Yet, as Hans Jonas *explained in "The Imperative of Responsibility, In Search of an Ethics in a Technological Age,"* because of the power of modern technology to create catastrophic harms such as those harms now foreseeable from human-induced climate change, ethics requires that policy-makers approach these matters with a "heuristics of fear," replacing the former "projections of hope" that traditionally guided policy (Jonas 1984, p. x). Yet, mainstream science is often uncomfortable with conclusions not grounded in scientifically verifiable observations. If this is so, ethics requires that IPCC's mandate be amended to synthesize scientifically plausible conclusions about climate change outcomes.

But to do this, the world needs institutions that will publish scientifically plausible impacts of human activities, a project which is not within the mission of most scientific research institutions. Also the international community needs scientific journals which will publish scientifically informed judgements of plausible adverse impacts from modern technologies.

For some potential environmental problems, scientific procedures have been developed which allow policymakers to make regulatory rules to protect human health from potential threats despite the fact that the threats have not been proven to cause harms to life or ecological systems. For instance, many governments have regulated potentially hazardous substances even though they have not proven to be harmful but are suspected of being toxic because they have properties similar to substances known to be toxic. Such an approach survives legal attack because many environmental laws don't require the government to prove some human activities are harmful but only require the government to demonstrate the government program is not "arbitrary and capricious." Governments can survive legal challenges to regulatory rules that limit some environmental threats by giving the public an opportunity to comment on proposed rules and responding to the comments. Such a process has often been sufficient to withstand legal challenges that the government's regulatory decisions have been "arbitrary and capricious."

Another way governments have effectively dealt with scientific uncertainty in environmental policy disputes is to shift the burden of proof to opponents of environmental policies to prove that a potentially dangerous human activity is safe. This has been the general approach of some European approaches to dealing with regulatory decisions that must be made in the face of scientific uncertainty (European Commission 2017). This approach is particularly appropriate as an ethical matter when a government is called upon to prevent extraordinarily dangerous potential harms from human activities. Following this logic, a reasonable response to those who resist climate change policies designed to protect human health or ecological systems on the basis of scientific uncertainty would be to shift the burden of proof to opponents of proposed climate change policies.

References

Anthony W et al (2018) 21st-Century modeled permafrost carbon emissions accelerated by abrupt thaw beneath lakes, https://www.nature.com/articles/s41467-018-05738-9. Accessed 14 Sept 2019

Bamber JL et al (2019) Ice sheet contributions to future sea-level rise from structured expert judgment. Proc Natl Acad Sci 116(23):11195–11200, https://www.pnas.org/content/116/23/11195. Accessed 12 Sept 2019

Breakthrough Institute, What Lies Beneath, (WLB) (2018) On the understatement of existential climate risk, https://docs.wixstatic.com/ugd/148cb0_a0d7c18a1bf64e698a9c8c8f18a42889.pdf. Accessed 21 Oct 2019

Brown D (2019) Six things that citizens around the world urgently need to know about climate change in light of several recent scientific reports, Ethicsandclimate.org. https://ethicsandclimate.org/2018/11/28/six-things-that-citizens-around-the-world-urgently-need-to-know-about-climate-change-in-light-of-several-recent-scientific-reports/. Accessed 22 Oct 2019

European Commission (2017) The Precautionary Principle: Decision-making Under Uncertainty

Intergovernmental Panel of Climate Change (IPCC) (2018) Special report on warming of 1.5 °C. https://www.ipcc.ch/sr15/. Accessed 8 Oct 2019

Intergovernmental Panel on Climate Change (IPCC) (2019) Summary for policymakers of IPCC special report on global warming of 1.5 °C. https://www.ipcc.ch/2018/10/08/summary-for-policymakers-of-ipcc-special-report-on-global-warming-of-1-5c-approved-by-governments/. Accessed 10 Oct 2019

Jonas H (1984) The imperative of responsibility; in search of an ethics for a technological age. University of Chicago Press, Chicago

National Geographic (2018) Climate change impacts worse than expected, global report warns, https://www.nationalgeographic.com/environment/2018/10/ipcc-report-climate-change-impacts-forests-emissions/. Accessed 8 Oct 2019

Nesbit EG et al (2019) Very strong atmospheric methane growth in the 4 Years 2014–2017: Implications for the Paris Agreement. Global Biogeochem Cycles 33(3):318–342. https://agupubs.onlinelibrary.wiley.com/doi/pdf/10.1029/2018GB006009. Accessed 9 Oct 2019

Stephan W et al (2018) Trajectories of the earth system in the anthropocene. Proc Natl Acad Sci 115(33):8252–8259, http://macroecointern.dk/pdf-reprints/Steffen_PNAS_2018.pdf. Assessed 30 Sept 2019

United Nations Framework Convention on Climate Change (UNFCCC) (1992) FCC/INFORMAL/84/Rev.1 GE.14-20481 (E)

World Resources Institute (2018) Half a degree and a world apart: the difference in climate impacts between 1.5 °C and 2 °C o f Warming. https://www.wri.org/blog/2018/10/half-degree-and-world-apart-difference-climate-impacts-between-15-c-and-2-c-warming. Accessed 14 Oct 2019

Animal Based Industries and Climate Change

Virginia Zambrano

Abstract The present chapter focuses on the relation between climate change and the animal-based industries. Firstly, the contribution of food production is analysed. It is argued that the agriculture impacts on climate change in a very profound way, not only by causing the loss of biodiversity, but also by polluting the land and the air, with severe damages for the health and the environment. Secondly, will be analysed the contribution of livestock and the animal-based industries to the climate change. It is argued that the livestock and the meat's industries have big economic interests in promoting meat consumption. This is why they credit the idea that eating meat is good for health. The aim of the present chapter is to stress the importance of a more responsive attitude towards meat's consumption. The needed changes claim for a collaboration amongst different actors, breeders, manifacturers, retailers, in order to minimize the environmental damages.

1 The Contribution of Food to Climate Change

The IPCC Special Report on the impacts of global warming of 1.5 °C above pre-industrial levels includes four illustrative pathways for mitigating GHG emissions, all of which explicitly separate agriculture, forestry and other land use from fossil fuels and industry. In spite of what is generally thought, one of the leading contributors to the environmental damage is agriculture. Agriculture impacts on climate change in a very profound way causing water pollution, and damages to land and biodiversity. According to Bellarby et al. (2008), agriculture is responsible for the release of carbon dioxide due to deforestation, methane from livestock and rice production, use of fertilizers (nitrous oxide) and fossil fuel energy, causing global warming. The use of fertilizers (as reported by the IPCC 2007) has proved, for

V. Zambrano (✉)
Capitol, Salerno, Italy
e-mail: vzambrano@unisa.it

L. Westra et al. (eds.), *Ecological Integrity in Science and Law*,
https://doi.org/10.1007/978-3-030-46259-8_2

example, to have negative effects on the environment, with significantly damages for the economy and the health (Reynolds and Nierenberg 2012).

The fertilizers, in fact, are responsible for the release into the atmosphere of toxic substances that, by spreading into the air, can also destroy or intoxicate organic crops cultivated miles away from the fields were the toxic substance has been originally employed. In this sense, if it is true that fertile land is essential for food production, it is also true that often some inappropriate agricultural practices are at the origins of the rapid declines in soil fertility and desertification. FAO studies indicate that in many countries, less than 10% of all spray applications hit a sick plant, a weed or a parasite, which means a waste of 90% of the product. The intensity of tillage and grazing, the use of fertilization and the application of pesticides (FAO 2012, table 15, p. 105) have reached critical levels in some geographical areas, where also the scarcity of water affects the production of food, calling for investments to promote a more efficient water management system.

No doubt that some technological innovations, such as the ones related to the so called Precision Agriculture (PA), can play an important role in reducing agriculture-related methane emissions. PA, in fact, is a modern management system using digital techniques to monitor and optimize agricultural production processes. It intends to measure variations in growing conditions within a field and to adapt the fertilizing or harvesting strategies accordingly. Likewise, it assesses the needs and conditions of individual animals in larger herds and optimizes feeding on a per-animal basis. But although the techniques of PA can serve to minimize the damages to the environment, there is no doubt that the consumption of land linked to the production of food and the breeding of livestock continues to grow at an unsustainable rate. The negative effects associated with this large-scale food production can also be appreciated in relation to the loss of biodiversity (Crist et al. 2017).

Although it is said that the loss of biodiversity is, to some extent, an unavoidable natural process, the contribution of humans to the extinction of species has improved recently by at least 100–1000 times the natural rate (FAO 2007). The need to reduce the variety of species used for food production and the expansion of genetic engineering, to increase the food production and to satisfy the economic interests of the manufactures and of the large multinational companies, has contributed to a progressive use of the herbicides and others chemical products for the pest resistance. In that way the loss of biodiversity endangers soil fertility and land productivity and is one of the prime factors contributing to the current 6th mass extinction (Geballos et al. 2017).

What has been defined as "a creeping degradation of the land and water systems that provide for global food security and rural livelihoods" (FAO 2011) is expected to worsen as a consequence of the climate change (IPCC 2012; Easterling et al. 2007). Though the agriculture is one of the key drivers for biodiversity losses, a leading source of this negative process is the livestock production, since the emissions from feed production generally exceed those of vegetable protein farming. A 2017 interesting study published in the journal Carbon Balance and Management—based on data from the Intergovernmental Panel on Climate Change (Wolf et al. 2017)—highlighted that animal agriculture's global methane emissions are 11% higher than previous estimates.

The types of gases produced from livestock production (the enteric emissions of the animals, the manure, the feed production) contribute in a significant way to climate change. Additional emissions are also created by animals due to the manure or the aquaculture ponds (Poore and Nemecek 2018). Nonetheless the same emissions from slaughterhouses and from processing meat industries, the wastage of fresh animal products contribute in a very significant way to pollution (Cooper et al. 2000).

2 The Impact on the Environment

On a closer inspection, not only impact the livestock and the meat transformation industries on climate change, they affect also the lives of people, living close to facilities where animals are kept confined and fed. In the past few decades, livestock farming has undergone a significant transformation, shifting from smaller, family-owned farms to large farms corporations. New technologies have allowed farmers to reduce costs, and increase profits, while the use of special feeds and pharmaceuticals products for animal have increased the efficiency and productivity of animal agriculture. The benefits for the market are evident as a large-scale industrial agricultural facility permits the raising of the animals, in much less time than before and is able to satisfy the growing consumption's demand of meat, eggs, or milk. The efficient feeding of animals increases the livestock facility size, and can positively impact on the local economy and lead to a permanent increase in employment (Schmalzried and Fallon 2007).

However, if this is true is also evident that these aforementioned advantages are mitigated by a series of health and environmental problems. One of the most pressing issues associated with this way of production stems from the amount of manures these industries produce. The manure, in fact, contains a variety of pollutants, such as chemicals used as additives to the manure, antibiotics, nitrogen, phosphorus, growth hormones, animal blood, pathogens such as E. coli etc. (Burkholder et al. 2007), which can contamite ground and source water. The quantity of manure production depends, of course, on the size of the farm, and creates major disposal problems as no sewage treatment plants exists for the livestock waste. The untreated manure is either spread on the ground or stored in clay or concrete pits under the buildings that hold animals, until it is used. Of course, the use of manure as fertilizer depends on the fact that the farming industries grow their own feed, which doesn't always happen. The large production and use of manure impact humans and led to growing environmental problems as it can affect the ground, the surface water quality and the ambient air quality, contributing to the global climate change. The environmental problems caused by these large industrial farms are therefore deeply connected with this intensive way of production. Where, as in the United States, this transformation has assumed considerable dimension, a growing body of legislations has been enacted trying to regulate this kind of activity.

In the USA, for example, the Concentrated Animal Feeding Operations (CAFOs) is an industrial agricultural facility where animals are raised for the production of meat, eggs, or milk. A CAFOs allows farmers to reduce costs, which results in greater profit and more economic incentives to increase farm size, but creates many environmental problems. Aware of these problems, the Environmental protection Agency (EPA), under the Clean Water Act (Sect. 502, CWA, 1972), than implemented a strict pollution control program—the National Pollutant Discharge Elimination System (NPDES)—setting the effluent limitation guidelines and standards (ELGs) for CAFOs. Under this regulation, farms industries should develop, and implement an adequate management plan for treating or processing their waste. Nevertheless, ruling in favor of farm producers, the judges of the Second Circuit Court of Appeals (in Water Keeper et al. vs. the EPA (2005) 399, F.3d 486) allowed alteration to the CAFOs permitting system. As a consequence of this decision, in October 2008, the Environmental Protection Agency (EPA) issued a new regulation to revise the rules governing the waste discharges from CAFOs. The current CAFO's rule eliminates the so called "duty to apply", i.e. the duty for the industries, either to apply for discharge permits or to demonstrate that they have no potential to discharge, and merely establishes the right of the environmental groups to access to the management plans. The same in Europe. The Dir. 91/676/EEC requires Member State the elaboration of Action Programs for the agronomic use of zootechnical waste (i.e. communication spreading and, if imposed, the Agronomic Utilization Plan). This Directive is still poorly respected due to the high number of intensive farms, the opposition of the producers and the few areas suitable for waste disposal.

Even in the presence of a regulation, no doubt that the spraying of liquid manure on local fields or its storage in ponds exposes the residents to adverse mental and physical health outcomes, harmful odors, diseases such as respiratory issues, nausea, increased blood pressure, asthma symptoms for children (Wing et al. 2012). The example coming from the hog farms, in USA, is paramount. In 2004 people living in North Caroline, near Concentrated Animal Feeding Operations (CAFOs), successfully persuaded the jury that industrial-scale hog operations created nuisances to their health and property. The overwhelming odors that cannot get out of the clothes, the toxic emissions and the swarms of insects entitled residents to bring an action in nuisance, in order to recover from the damages suffered. In Gillis *v.* Murphy-Brown, LLC (E.D.N.C., No. 14-cv-00185, jury verdict 12/12/18), the judges than recognized that Smithfield Foods was liable for the damages caused by the high-density livestock operations, but awarded to the nine plaintiffs only compensatory damages. The dismissal of punitive damages claim was based on the fact that, under North Carolina law, punitive damages only serve "to punish a defendant for egregiously wrongful acts and to deter the defendant and others from committing similar wrongful acts" (N.C. Gen. Stat. § 1D-15). In the case at issue, the judges found that the defendant could only be held liable for compensatory damages, on the evidence of a willful or negligent conduct. The same limitations exist under Georgia Law, whose standard for allowing punitive damages implies the evidence of a "willful misconduct, malice, fraud, wantonness, oppression, or that entire want of care which would raise the presumption of conscious indifference to consequences" (O.C.G.A. § 51-12-5.1(b).

On the one hand, is worth noting that the Smithfield Foods defense was based on the false assumption that the measures to reduce the odor were adequate, the claim exaggerate and filed by out-of-state lawyers who went door to door, "actively recruiting plaintiffs and promising them large sums of money if they joined the lawsuits". On the other, the recognition of the right to farm has often influenced the law suit in favor of the agrobusiness industries (Carrie Hribar 2010). Relying on this line of reasoning, the judges, though recognizing the damages, have reduced their amount pro capita. The tendency to prioritizes private interests rather than the well-being of communities and the protection of the environment is evident and can be also inferred by the ambiguous role played by the EPA. The environmental and public health damages caused by CAFOs, is also associated to some others negative socio-economic consequences.

The meat production and distribution systems are controlled by large meat processing companies whose strategy, starting from the beginning of the '80, was to abandon the old-line plants in the cities and moving to rural communities. In choosing cheapest production sites, those agrobusiness multinational corporations cut their labor costs by creating low-wage and deskilled jobs offered to minorities, immigrants, refugees, and women, mostly coming from Mexico. Nowadays, the so called "meatpacking communities" have become in Dakotas, Minnesota, Nebraska and Iowa, Kansas, Northern Texas and Missouri, a new rural reality where thousands of poor people live, with few chances of getting a professional qualification and improving their life conditions (Eastwood et al 2010). More. Tyson, Pilgrim's, Smithfield, SaraLee, ADM Cargill, National Beef, Conagra—just to name few of the big agrobusiness corporations—are able to vertically control every aspect of the meat production, from the ownership of the land to the processing of meat, its distribution, and sales. While in the '70 the top four beef packing companies slaughtered only the 29% of steers and heifers, today this percentage has risen to over 80%. By reducing the costs, not only are capable these scale productions to better respond to the increasing food demand, they also enable the corporations to impose a number of economic barriers to small producers. The difficulties in accessing to land and to credit, the lack of competitiveness and the rising of the risk factors push the smallholder productors down to a semi-subsistence level rate, forcing them to drive out of the market (Welsh 1996). The CAFOs offer a good example of the noumerous problems raised by the animal-based industries.

3 Meat Consumption and Market Strategy

Despite all these problems caused by the meat production—not least the fact that the World Health Organization in 2015 categorized red meat as carcinogenic to humans and a possible source of others non-transmissible diseases (obesity, diabetes)—the meat consumption has doubled in the last 50 years, as result of the population growth and the rising of individual income. This is particularly true for the Middle-income countries, as China or east Asia that, by the mid-past century, are experiencing a 76% increase in the consumption of poultry, a 69% increase in beef and a 42% increase in pork. This situation is counterbalanced by a declining trend in the developed countries where a higher education level, a more conscientious attitude toward the environment and the health have led to a significant decrease in the consumption of raw beef, lamb and pork that, since 2012—as reported by the UK's 2017 National Food Survey— dropped more than 4.2% as well as that of other meat products including sausages, bacon and poultry. Despite this dropping of the meat consumption in the developed countries, the Europe as well as the USA are experiencing an overall growth in size of the livestock breeding facilities, due to the increasing values of the economic activities related to the meat processing. In this context, the meat corporations are driving the consumer demand by way of aggressive advertising campaigns aimed at create in the consumer the need for meat.

This is particularly evident in USA, where the meat industries rely on marketing techniques to "produce meat's demand" and to influence and shape the people's food choices. To say that the consumers' habits—especially those of children perceived as future consumers—are "controlled" by the market strategies. On the one hand, the idea that eating meat gives more strength and that the meat is full of proteins and minerals, essential in every phase of the life, from pregnancy to old age, is used to "naturalize" people's habits. On the other, the meat is represented as a healthy, natural, not processed, food compared to the ordinary junk food. The same in Europe, where some meat producers associations (Assocarni in Italy, Interveb in France, VLAM in Belgium, SBK in Holland) have launched advertising campaigns in defense of the livestock sector and of the industrial, artisanal and commercial meat processing activities, to demonstrate the importance of meat consumption in the diet. All the adverts' campaigns dedicated to the products of animal origins (meat, cheese, milk, eggs etc.) not only show that animals are cared for, they also highlight—contrary to the truth—that only meat can satisfy the human need for proteins. It is clear, therefore, that the consumers' education and their food choices can play an important role in reducing the consumption of natural resources for food production.

4 The "Tobacco Litigation" as a Possible Model for Future Suits Against Animal-Based Food Industries

The documented link between the industrial livestock production and the outbreaks of illnesses such as heart disease, obesity, diabetes and certain types of cancers (Pan et al. 2012), not only should led to further the awareness that consuming animal products is dangerous for the health, but also to an increased regulation of animal-based food industries legislation caring for the consumer health (Winalski 2013).

The *Hallmark Meat Packing case* offer a good example for future suits against animal-based food industries and illustrated how the meat industries try to conceal harmful facts. Inspecting a *Hallmark* plant in California, in 2007, the Human Society of the United States (HSUS) found that behind the plant's doors, animals were treated and slaughtered with cruelty (Perry and Brandt 2008). After having the USDA (United States Departement of Agriculture) and the Food Safety and Inspection Service (FSIS) completed their inspection, Hallmark was issued a Notice of Suspension for "failure to maintain and implement controls to prevent the inhumane handling and slaughter of animals" (USDA 2008) and its contract was rescinded.

As a consequence of that, the HSUS filed a lawsuit against Hallmark for deceiving the Agricultural Marketing Service (AMS). The action was brought under the *qui tam* (whistleblower) section of the False Claims Act. At issue was the liability of *Hallmark* for represented falsely to the AMS that all cattle slaughtered were handled humanely, in accordance with federal regulations, and no ill animal was included in the AMS purchases. Nevertheless, the core of the lawsuit was not based on the bad animals' treatment, but on the fact that the corporation committed fraud by representing (contrary to the truth) that they treat cattle humanely. The whistleblower, by revealing the cruelty against the animals, presented evidence that the corporation acted in direct violation of the contract with the USDA, whose action was than based on the need to recuperate the taxpayers' money spent on recalled beef (Lovvorn and Perry 2009).

In the case of *Hallmark* the factors that helped in winning the claim were a successful allegation of fraud, made possible under the False Claim Act (FCA 31 U.S.C. §§ 3729 to 3733), the intervention of the government as plaintiff, the central role played in the claim by the government's interests in protecting the health of the children.

Infact, in spite of the message that eating meat is necessary for a healthy diet, meat industries (and *Hallmark* in this case) are well aware of the damages caused to the health by the industrial livestock production.

In *Hallmark*, although the contract did not clearly involved the supply of healthy food, it implicitly required the food to be safe, especially because it was food that would have been eaten by young children going to school. It was on these assumptions that the USDA rescinded the contract and condemned *Hallmark* to pay for damages.

Some scholars (Winalski 2013) have drawn a parallel with the Tobacco Litigation cases, underlying the subtle contradiction between a policy oriented, on the one hand, at warning about the dangers of smoking but encouraging, on the other, the consumption of milk, dairy, beef etc. that may be harmful to health. In the case of the Tobacco Litigation, as it is well known, the problems for plaintiff was to meet the burdens of proof on causation, injury, and damages.

The identification of the proximate cause of injury required relevant expertises and the need of engaging in a legal battle, that only deep-pockets subjects could afford. The problems in proving the foreseeability of the damage, as well as the defense represented by the "voluntary conduct of the consumer", at first, led the courts to rule in favor of the tobacco companies (Gifford 2010). But if in U.S. v. Phillip Morris, Inc., 116 F. Supp. 2d 131 (D.D.C. 2000) the judges were still not convinced of the plaintiffs' position, some years later they changed their opinion and, in U.S. v. Philip Morris USA, Inc., 449 F. Supp. 2d 1, 920–21 (D.D.C. 2006), the court held the defendants was even liable for racketeering. Finally, in 2009, the government with the adoption of the Family Smoking Prevention and Tobacco Control Act (FSMTCA 2009), imposed restrictions on the advertising and promoting of the cigarettes.

The Tobacco Litigation offers an instructive example of how making the animal-based food industries liable for the harms they cause by supplying foods that —as supported by medical literature and science—are detrimental to the public health.

In the *Hallmark* case, the contract dealt with the supply of animal-based food to schools. The company was, then, under the obligation to inform the consumers of the possible risks connected to the use of products which could have detrimental effects on health and on the environment.

In *Hallmark*, the disclosure of the whistleblower made clear that not only the company knew that its conduct was in violation of the federal contracts but that, by intentionally concealing such informations, this conduct amounted to fraud. The conduct of *Hallmark* was even more harmful. The potential victims of the fraudulent conduct were children, eating at school the food provided under the National School Lunch Program. Here, the familiar refrain of "personal responsibility" as the traditional defense used by the meat corporations, when discussing food habits couldn't work. In this case, the harm was primarily caused to children's health and then to the government for its exposure to the health's care expenses. In the case of the so-called Tobacco's Litigation the focus on the harm caused to children led—*inter alia*—to the adoption of the Family Smoking Prevention and Tobacco Control Act (2009) trying to impose some restrictions on Tobacco's industries to prevent the spreading of smoke among younger.

Even so, despite the warnings, no doubt that the financial interests of the big smoking corporations continue to affect political choices. The same for the animal-based industries whose economic power charges the public authority with the impossible task of both promoting the health of its citizens and promoting the consumption of meat.

5 A New Dietary Culture

To avoid the problems linked to meat production and consumption, a possible solution could be met in the transformation of the food production relationship, in a more responsive attitude towards food consumption and in the use of sustainable practice for food manufacturers, retailers, and food services (Ripple William et al. 2017). Land conversion should be avoided by developing best practices, including limited fertilization, cover crops, crop rotation, organic production, well-managed grazing, for example, animals graze in forested areas. Lowering the use of land- and GHG-intensity emissions connected to the food consumption implies also a shift in dietary choices, favouring the use of foods with lower emissions and a reasonable consumption of land. As it has been said "a food system in which quality farmland is devoted to raising cattle feed instead of crops is a good way to waste resources, impoverish farmers and maintain hunger" (Lappe 1971). The transition to sustainability implies the adoption of several policy instruments, able to influence the markets and other economic drivers. The steps towards an alternative way of doing business include the need (1) to a well-managed reserves of the world's terrestrial, marine, freshwater, and aerial habitats for maintaining nature's ecosystem; (2) to halt the conversion of forests or grasslands; (3) to reduce food waste through education; (4) to encourage dietary shifts towards plant-based foods; (5) to promote new green technologies and the use of renewable energy sources; (6) to reduce wealth inequality and to improve workers' condition; (7) to adopt economic policies able to countering the interests of the large multinationals operating in the meat sector.

As if to say that, if the market directs economic choices, then only a progressive change in alimentary habits can orient towards a sustainable agriculture. The message, to motivate the consumers to change their diets, should than consider all factors which influence consumer behavior towards food. The aim is to convince the public that the consumption of meat is neither natural nor normal and doesn't meet the protein needs of the human body, as generally believed. The adverts such as "eat a juicy steak," or "Beef¾Real Food for Real People" are aimed at influencing the choices of consumers, especially children and men, who are more sensitive to the idea that eating meat strengthens the body. Therefore, the large multinationals operating in the meat sector are interested in promoting and encouraging higher levels of meat consumption.

Well-developed and attractive new meat alternatives should than be used together with social marketing techniques to convey the need for dietary changes and to promote a shift in dietary habits. The need to encourage a shift in dietary habits was, for example, at the origin of the American Meatless Monday's campaign, aimed at supporting health, nutrition, animal welfare and the environment. Nonetheless this change, although desirable, is not easy to achieve, due to the strong economic interests of the meat-producing industries. The reaction of the National Cattlemen's Beef Association to the campaign launched by the U.S. Department of Agriculture (USDA) in 2012 is emblematic. The USDA, in "Greening Headquarters Update", emphasized the contribution to greenhouse gas emissions and health risks of the animal agriculture, inviting its employees to join the Meatless Monday's campaign. The immediate protest of the National Cattlemen's Beef Association and of the beef-producing states forced the USDA to tweet a statement for removing the endorsement to the Meatless Monday. One of the reasons was that this campaign, used during the World War I against an excessive meat consumption (Wilson 1918) couldn't be endorsed by a public agency and, therefore, employed as a weapon against producers whose primary aim was to produce food and fiber for a growing global population in a very sustainable way. What lesson can we draw from all these events? It emerges that the central question is not whether to eat or not meat, but how to reconcile the opposing interests. At issue is not the possibility of imagining a vegetarian world, but the hope of focusing on a more plant-based diet. Suitable institutional and policy frameworks should than be developed, at local, national and international levels, to increase in the consumers the awareness that continuing "business as usual" is detrimental for the health and the environment, and to show the benefits of a changing attitude (Fox 1991).

References

Bellarby J et al (2008) Cool Farmin: Climate Impacts of Agriculture and Mitigation Potential, University of Aberdeen, 2.1 (13)

Burkholder J, Libra B, Weyer P, Heathcote S, Kolpin D, Thorne P et al (2007) Impacts of waste from concentrated animal feeding operations on water quality. Environ Health Perspect 11(2):308–312. Available at http://www.ncbi.nlm.nih.gov/pmc/articles/PMC1817674/pdf/

Carrie Hribar MA (2010) Understanding Concentrated Animal Feeding Operations and Their Impact on Communities, Mark Schultz, National Association of Local Boards of Health, www.nalboh.org, p 11
Cooper M, Rosset P, Bryson J (2000) WARNING: corporate meat and poultry may be hazardous to workers, farmers, the environment and your health. Reimagine! 7(2):30–33. Available at https://www.jstor.org/stable/41554280. Accessed 27 Mar 2019
Crist E, Mora C, Engelman R (2017) The interaction of human population, food production, and biodiversity protection. Science 356:260–264
Eastwood R, Lipton M, Newell A (2010) Farm size. In: Pingali P, Evenson R (eds) Handbook of agricultural economics, vol 4. Elsevier, Amsterdam, pp 3323–3397
Easterling D et al (2007), Food, fiber and forest products. Climate change 2007: impacts, adaptation and vulnerability, Cambridge University Press, 272–313
Family Smoking Prevention and Tobacco Control Act (FSMTCA 2009) 21 U.S.C. § 387 (2012) Available at http://www.gpo.gov/fdsys/pkg/USCODE-2012-title21/pdf/USCODE-2012-title21-chap9-subchapIX.pdf
FAO (2011) The State of the World's Land and Water Resources for Food and Agriculture. Managing systems at risk, FAO Conference document C2011/32. Thirty-seventh Session. Rome, 25 June–2 July. Available at: www.fao.org/nr/solaw/solaw-home/en/
FAO (2007) and (2012) The State of Food and Agriculture Report. Available at www.fao.org/3/i3028e/i3028e00.htm
Fox MW (1991) The Place of Farm Animals in Humane Sustainable Agriculture. Humane Society USA, Washington, DC, 1991. See, also, The Human Society of the United States (2010) Factory Farming in America. The True Cost of Animal Agribusiness for Rural Communities, Public Health, Families, Farmers, the Environment, and Animals. Available at https://www.humanesociety.org/sites/default/files/docs/factory-farming-in-america-true-cost.pdf
Geballos G et al (2017) Biological annihilation via the ongoing sixth mass extinction signaled by vertebrate population losses and declines. Available at http://www.pnas.org/content/114/30/E6089
Gifford DG (2010) Suing the Tobacco and lead pigment industries: government litigation as public health prescription. University of Michigan, pp 36–37
IPCC (2007) Fourth Assessment Report. Available at http://www.ipcc.ch/assessment-report/ar4/
IPCC (2012) Managing the risks of extreme events and disaster to advance climate change adaptations. Available at http://www.ipcc.ch/site/assets/uploads/2018/03/SREX_Full_Report-1.pdf
Lappe FM (1971) Diet for a small planet. Ballantine Books, New York
Lovvorn JR, Perry NV (2009) California proposition 2: a watershed moment for animal law. Animal Law 15:149, 156
Pan A et al (2012) Red meat consumption and mortality. Arch Int Med 172:555–563
Perry N, Brandt P (2008) A case study on cruelty to farm animals: lessons learned from the hallmark meat packing case. Mich Law Rev First Impressions 106:117, 119. Available at http://www.michiganlawreview.org/assets/fi/106/perrybrandt.pdf
Poore J, Nemecek T (2018) Reducing food's environmental impacts through producers and consumers. Science 360(6392):987–992. https://doi.org/10.1126/science.aaq0216
Reynolds L, Nierenberg D (2012), Innovations in sustainable agriculture: supporting climate-friendly food production, washington D.C.
Ripple William J, Wolf C, Newsome TM, Galetti M, Alamgir M, Crist E, Mahmoud MI, Laurance WF (2017) 15,364 scientist signatories from 184 countries. BioScience 67, 12:1026–1028. https://doi.org/10.1093/biosci/bix125
Schmalzried HD, Fallon LF Jr (2007) Large-scale dairy operations: assessing concerns of neighbors about quality-of-life issues. J Dairy Sci 90(4):2047–2051
USDA (2008) Audit Report: Evaluation of FSIS Management Controls over Pre-Slaughter Activities, Rpt. No. 24601-7-KC, i. Available at http://www.usda.gov/oig/webdocs/24601-07-KC.pdf

Welsh R (1996) The Industrial Reorganization of U.S. Agriculture: An Overview & Back- ground Report. In: Henry A. Wallace Institute for Alternative Agriculture Policy Studies Report No. 6, Greenbelt, MD
Wilson W (1918) President's proclamation calling upon people of nation to reduce consumption of wheat and meat products in order to feed America's associates in the War. In: Rochester ES (ed) Official Bulletin. 2 (219). See also: The Monday Campaigns, Meatless Monday: History, http://www.meatlessmonday.com/about-us/history. Available at http://perma.cc/02mVzNJgwHs] (accessed June 17, 2019)
Winalski A (2013) Shocked, horrified, sickened: how cigarettes (and the lessons from the tobacco litigation) can take years off animal-based food industries. Animal Law 20(167):171
Wing S, Horton RA, Rose KM (2012) Air pollution from industrial swine operations and blood pressure of neighboring residents. Environ Health Prospect. https://doi.org/10.1289/ehp.1205109
Wolf J, Ghassem RA, Tristram OW (2017) Revised methane emissions factors and spatially distributed annual carbon fluxes for global livestock. Carbon Balance Manag 12:16. Available at. https://doi.org/10.1186/s13021-017-0084-y

Climate Science Before the Courts: Turning the Tide in Climate Change Litigation

Matteo Fermeglia

Abstract This chapter aims at shedding a light on the pivotal role played by climate science within the ramping climate justice debate. By examining three different cases held at different latitudes, it wishes to explain how climate science has been successfully employed in climate litigation to establish State's responsibility for more ambitious policies; to determine bans on carbon-intensive infrastructures; to establish patterns of liability for climate-related damages. In this regard, this contribution champions climate science as a fundamental stepping-stone towards judicial law-making amidst the climate crisis.

1 Introduction

Courts are increasingly opening the gate to climate change. As of 2017, more than 850 climate change cases had been filed in 24 countries (including the European Union), which altogether contribute to what has now been widely acknowledged as a specific and self-standing branch of litigation, namely *climate change litigation* (UNEP 2017). More specifically, the notion of climate change litigation encompasses "any piece of federal, state, tribal, or local administrative or judicial litigation in which the [...] tribunal decisions directly and expressly raise an issue of fact or law regarding the substance or policy of climate change causes and impacts" (Markell and Ruhl 2012, p. 27). Within such conceptual framework, three main categories of claims can be identified, which will also operate as a blueprint for the present inquiry:

- Litigation brought against States (or public authorities) for failing to adequately tackle climate change (*pro-regulatory litigation*);

M. Fermeglia (✉)
Hasselt University, Faculty of Law, Hasselt, Belgium
e-mail: matteo.fermeglia@uhasselt.be

L. Westra et al. (eds.), *Ecological Integrity in Science and Law*,
https://doi.org/10.1007/978-3-030-46259-8_3

- Litigation brought before public bodies to ban specific energy-intensive projects (*permitting litigation*);
- Litigation brought against private entities to seek remediation for climate-related damages (*liability litigation*).

Common hallmark to all these patterns of litigation is its close intertwining with the long-standing deal of findings provided by climate science. Sound scientific evidence has clearly enhanced social and institutional awareness about the inadequacy of the current system—at the international, as well as national level—to ensure a meaningful reduction of GHG emissions by major emitting countries and companies. Hence, while fostering certainty as to the actual anthropogenic contribution to climate change, climate science spurred the outbreak of lawsuits mainly directed at holding public entities accountable for their climate commitments (Markell and Ruhl 2012). Moreover, the scientific knowledge about specific attribution of historical—as well as present—GHG emissions worldwide is rapidly gaining ground. Therefore, the target of climate litigation is shifting towards holding private emitters responsible *vis-à-vis* civil society for knowingly jeopardizing the climate through their economic activities (Duffy 2009).

As meritorious and bold such claims may be, they have faced several setbacks due to manifold legal barriers arising in different legal and political contexts. Against this backdrop, this contribution aims to strengthen the view that strategic use of climate science may substantially contribute to a deference shift in the judicial approach towards climate change claims.

This contribution goes as follows. Section 2 highlights the most relevant subsets of climate science as relevant for the selected climate litigation cases. Section 3 analyses how such scientific evidence has been successfully invoked in three climate cases belonging to each of the above branches of climate litigation: *Urgenda Foundation v. The Netherlands* (pro-regulatory litigation); *Gloucester Resources v. Minister for Planning* (permitting litigation); *Lliuya v. RWE* (liability litigation). Building on the case studies, Sect. 4 briefly concludes yielding the main argument of this chapter, namely that climate science holds a prominent role in the uptake of judicial rule-making in the climate arena.

2 What Climate Science in Climate Litigation?

From a climate science perspective, detection of human caused climate change and attribution of climate change-related events are scientific terms that provide tools for the lawyer's task of showing the existence, causes and effects of climate change (Allen 2011; French and Pontin 2016, p. 11). Oddly enough, however, a little effort has been devoted thus far to a more qualitative appraisal as to *how* climate science indeed plays out in climate litigation (Setzer and Vanhala 2019, p. 580). Yet, while climate change lawyers need not be scientists, they need to understand the application of science, in terms of its uses and limits. For the purposes of this contribution, three subsets of climate science acquire relevance as presented in Court: (1) GHG *emission mitigation pathways*; (2) *carbon budget* calculations; (3) *attribution* of GHG emissions to specific entities. Relevant, such findings are mostly contained in IPCC's official documents, being the same IPCC "the global clearing house of climate research" (Ghaleigh 2016).

2.1 GHG Emission Pathways

Ever since its foundation in 1987, IPCC has extensively established through computer models (with a level of confidence of 9 chances out of 10) that a direct, linear link exists between anthropogenic GHG emission and average temperature rise (Mastrandrea and Schneider 2014). In this regard, since the drafting of the 1997 UNFCCC Kyoto Protocol targets, country-based elaborations on GHG mitigation pathways have been increasingly refined, up to the very specific long-term scenarios in the 2018 IPCC 1.5° Report. The 1.5° Report sets down the emission reductions necessary to ensure a 66% likelihood of achieving the temperature target agreed upon in the 2015 Paris Agreement. Accordingly, in model pathways with no or limited overshoot of 1.5 °C, global net anthropogenic CO_2 emissions must decline by about 45% from 2010 levels by 2030, reaching net zero around 2050 (Rogelj et al. 2018, p. 95). To limit global warming to below 2 °C, CO_2 emissions are projected to decline by about 20% by 2030 in most pathways, and eventually reach net zero around 2075 (Victor et al. 2014, p. 124). Non-CO_2 GHG emissions (CH4, sulphur dioxide, etc.) should follow similar emission reduction pathways (Rogelj et al. 2018, p. 96). Moreover, and perhaps more importantly, each IPCC global warming scenario is linked to specific implications in terms of climate-related risks to health, livelihoods, human security, and growth (Hoegh-Guldberg et al. 2018). Further calculations are annually conducted by UNEP to keep track of the progress made in achieving the above emission reduction pathways, thus highlighting the existing gap between the ideal trajectories and the actual collective efforts being made by countries (UNEP 2018).

Relevant from a legal standpoint, scientific findings on climate mitigation targets and pathways as factored into climate change scenarios are now capable of adequately appraising the short and long-term risks climate change will pose in terms of livelihood losses, damages to specific areas, impact on food production, and therefore in terms of disruption of property, family life, etc., thus pitching the proposed emission mitigation targets accordingly.

2.2 Carbon Budget

Explicit GHG emission reduction targets as set out in national (or supranational) legislation and policies may well be of assistance to courts seeking a defensible basis for 'scaling down' an assessment of the significance of a given project's climate change contribution. Yet, climate change is the paradigmatic *global* environmental problem, as anthropogenic GHG emissions are produced in all countries by innumerable entities. No one country or entity can be deemed to be *the* cause of climate change. Nonetheless, some countries and corporations are more substantial contributors to atmospheric GHG emissions than others.

This acquires peculiar relevance in the context of climate change lawsuits brought under various permitting legislation (e.g., Environmental Impact Assessment regulations), inasmuch as such pieces of legislation somewhat encompass GHG emissions—or, more generally, climate change aspects—as a significant impact of specific projects. In fact, in this field of litigation defendants often invoke the "drop in the ocean" argument, whereby GHG emissions stemming from a particular activity cannot be said to cause climate change harm and/or have a significant environmental impact in global terms, and therefore should be negligible for the purposes of the given permitting or planning procedure.

Against this backdrop, the existing *carbon budgets* calculations aim to unfold the cumulative amount of CO_2 emissions permitted over a period of time to avoid overshoot of the "well below 2°C" above pre-industrial levels overshoot, consistently with the objectives agreed globally and officially endorsed in the 2015 UNFCCC Paris Agreement.[1]

In sum, relying on Integrated Assessment Models or Earth Models, CO_2 budgets draw from life cycle emission estimates, which fundamentally determine the anthropogenic CO_2 input to the atmosphere as balanced by output in the carbon reservoirs on land or in the ocean, while taking into account also other factors beyond combustion—e.g., the differing degrees of carbon capture and storage (CCS) deployment.

Several carbon budgets have been published over the last decade. The most authoritative carbon budgets come as a result of scenarios developed in international policy documents, such as the IEA's World Energy Outlook, IPCC's Assessment Reports, as well as independent academic endeavours, such as the Global Carbon Project (Le Queré et al. 2018).[2]

According to IPCC's carbon budget calculations contained in the 1.5° Report, to ensure a greater than 66% probability of limiting global average temperature rise to no more than 2 °C, cumulative human CO_2 emissions since 1870 through year 2100 must be less than 1000 Gt (Rogelj et al. 2018, p. 104). If non-CO_2 greenhouse gases

[1] Relevant, specifying the level of 'pre-industrial emissions' aims to provide a baseline from which anthropogenic activity began influencing greenhouse gas (GHG) concentrations in the atmosphere. How the 'pre-industrial level' is interpreted, however, can vary. To date, 1850–1900 has been the preferred baseline by institutions including the IPCC. However, some studies have suggested a 1720–1800 baseline would be more appropriate because GHG concentrations have been increasing since industrialisation began around 1750. Others argue that baselines should be taken from natural climate model simulations, i.e. those that exclude anthropogenic forces.

[2] It is also worth noting that carbon budget calculations differ in terms of scope and range. For example, IEA's carbon budget refers to budgets for the energy sector only—the largest single source of CO_2 emissions through the burning of coal, oil and gas. In contrast, the IPCC's budgets account for all anthropogenic sources of CO_2, thus including, inter alia, budgets for heavy industries and land use, land use change and forestry (LULUCF).

are not reduced at the same rate, the carbon budget should be reduced by up to a further 210–790 Gt (Rogelj et al. 2018, p. 104). Cumulative emissions from 1870 through 2017 equalled 575 Gt (Collins et al. 2013; Le Queré et al. 2018). Thus, the remaining carbon budget amounts to 215 Gt. Given the current global annual emission rate of 10 Gt CO_2, the remaining carbon budget would be consumed in little more than two decades (Le Queré et al. 2018).

The most relevant consequence of carbon budgets evidence from a legal and policy relate to the need to halt further carbon intensive projects—such as greenfield oil sands, coal mines, etc.—over the next decades. A recent study commissioned by the European Union notes that such effort would be worth a 1.6 USD trillion risk on the oil and gas industry (Carbon Tracker Initiative 2018).

2.3 *Attribution of Historic GHG Emissions to Specific Entities*

The larger-than-average contribution to GHG emissions generated by some countries and companies, raises the issue as to whether such entities can be held individually responsible for the resulting climate change impacts, even where these effects will be felt globally.

Since 2013, scientists have begun to attribute operational and product GHG emissions, thus showcasing a producer-side view of climate accountability (Heede 2013). The Carbon Major Report now attributes 63% of the CO_2 and methane emitted between 1751 and 2010 to 90 entities: Fifty investor-owned companies, such as Chevron, Peabody, Shell, and BHP Billiton; thirty-one state-owned companies, such as Saudi Aramco and Statoil; nine government-run industries in countries, such as China, Poland, and the former Soviet Union (Heede et al. 2014).[3] The research also classified the 90 entities according to type of fossil fuel extracted and marketed—56 oil and natural gas companies and 37 coal producers. In addition, CO_2 emissions from seven cement manufacturers are included. Half of the emissions traced to the ninety "carbon majors" have occurred since 1986, demonstrating the increasing speed with which fossil fuels are being burned. These findings are far from settling the controversial moral issue as to who should bear present remedial responsibility as a result of historic GHG emissions (Butt 2017). Yet, they indeed constitute a path-breaking scientific development with thorough legal implications, as they enable to strengthen the causal chain between some specific entities' conducts and the—already recognized—hazardous risk of detrimental consequences of global warming.

[3] In sum, the calculations draw from companies or entities net fossil fuel production data from publicly available sources, as factored with each fuel's carbon content, deduction for non-energy uses of produced fuels, and emission factors for each fuel, for each entity, and for every year for which production data have been found.

3 Climate Science in Courtrooms

3.1 Urgenda Foundation v. The Kingdom of the Netherlands

The landmark *Urgenda* case certainly stands out as the first successful attempt to compel a State to enhance its climate policy's ambition, in order to align the latter with best climate science.[4] A Dutch environmental NGO (*Urgenda Foundation*) and almost 900 Dutch citizens brought a lawsuit against the Dutch State before the District Court in The Hague. Urgenda alleged that the State had violated its constitutionally imposed duty of care *vis-à-vis* their citizens, because the existing GHG emissions reduction pledge of 17% by 2020 as compared to 1990 levels would be inadequate to meet the State's fair contribution toward the goal of achieving global temperature increase within 2 °C of pre-industrial conditions.[5] According to Urgenda, the overall State's GHG emissions reduction threshold should amount to 25–40% below 1990 levels by 2020, consistently with the targets set for UNFCCC Annex I Countries. This benchmark is also consistent with IPCC's 450 ppm concentration scenario as designed in AR4 and AR5, which would allow a 50% change to remain below the 2 °C pivotal global warming threshold.

Relevant, IPCC's findings in AR4 (2007) and AR5 (2013) have been largely relied upon by Urgenda, and eventually understood by the Court, as *risk regulation tools*, since they "determine the desired convention and possible adverse effects" of climate change (para. 4.19). The District Court recognized the widespread consensus both in climate science and at the international climate policy level on the need to prevent certain detrimental consequences of climate change (para. 4.71). In view of risk management and based on IPCC's scientific evidence, utmost preference should be given to the 450 ppm global warming scenario, which entails a lower risk to face hazardous climate change (para. 4.22). And, by expressly endorsing the 2 °C target and the relevant scenarios formulated by the IPCC, States parties to the UNFCCC—including The Netherlands and the EU—have explicitly embraced the same 450 ppm scenario as "the starting point for the development of climate policies" (para. 4.14). Thus, the 25–40% GHG emission reduction target by 2020 set for UNFCCC Annex 1 Countries to achieve the 450 ppm scenario, was deemed by the Court as the threshold to establish Dutch State's duty of care to "make an adequate [mitigation] contribution, greater than its current contribution, to prevent hazardous climate change" (par. 4.89).

Importantly, the reasoning on scientific evidence, GHG emission reduction targets and trajectories underpins the reasoning on relevant legal tools to establish the State's duty of care *vis-à-vis* its citizens, namely: (1) Article 21 of the Dutch Constitution (On the duty of the State to protect and improve the environment); (2)

4 Urgenda Foundation v. The Netherlands [2015] HAZA C/09/00456689 (June 24, 2015).

5 Notably, this GHG reduction target is more ambitious than those agreed at the EU level, as disaggregated among Member States—including The Netherlands, which indeed amounts to 16% by 2020.

EU GHG emissions reduction targets; (3) the rights to life and to respect for private and family life as protected by the European Convention on Human Rights (Articles 2 and 8); (4) the no harm principle of international law, as laid down in the Trail Smelter case; (5) the doctrine of hazardous negligence under Dutch Civil Code (Section 6:162); (6) the precautionary and sustainability principle embodied in the UNFCCC; (7) the principle of a high protection level, the precautionary principle, and the prevention principle as embodied in the main EU environmental and climate policies (art. 191 TFEU).[6]

Climate science played an even more prominent role in the appellate judgment before The Hague Court of Appeal.[7] Whilst the Dutch government in the wake of the District Court decision adopted a 49% GHG emissions reduction target by 2030 (as compared to 1990 levels), it did not commit to an interim target of at least 25% reduction by 2020, thus limiting to the previous 20% reduction commitment as agreed at EU level.

The Appellate Court upheld the District Court decision, hence further recognizing the inadequacy of the interim 20% GHG emission reduction target to 2020 to fulfill the State's obligation to protect its citizens. To this end, the Court extensively relied on IPCC's AR5 to acknowledge: (a) the existing linear relation between GHG emissions and global warming[8]; (b) the general consensus on the scientific evidence that global temperature increase should not exceed 2° by 2100 (paras. 3.3–3.5); (c) that IPCC's AR4 and AR5 as officially adopted throughout all subsequent UNFCCC COPs, although "may not have established a legal standard with a direct effect", yet provide strong evidence of a "real threat of dangerous climate change, resulting in the serious risk that the current generation of citizens will be confronted with loss of life and/or a disruption of family life […]", which the State's ought to protect pursuant to Articles 2 and 8 ECHR (par. 45). Furthermore, the Appellate Court concluded that Dutch State's 2020 20% GHG emission reduction target had no substantiation based on climate science, while it was an established fact that postponing higher interim reductions would cause continued emissions of CO_2, which in turn contribute to further global warming (para. 52). More specifically, the Dutch State

[6] In particular, where determining the scope of the duty of care of the State, the court has taken account of:

(i) the nature and extent of the damage ensuing from climate change;
(ii) the knowledge and foreseeability of this damage;
(iii) the chance that hazardous climate change will occur;
(iv) the nature of the acts (or omissions) of the State;
(v) the onerousness of taking precautionary measures;
(vi) the discretion of the State to execute its public duties, with due regard to the latest scientific science, the available (technical) options to address necessary security measures to, and the cost-benefit ratio of the same security measures.

[7] Urgenda Foundation v. The Netherlands [2018] HAZA C/09/456689 (October 9, 2018).

[8] For example, in the discussion of the risk of reaching "tipping points" with a temperature rise of between 1 and 2 °C, the court cited the IPCC AR5 Working Groups I, II and III Synthesis Reports (para. 44).

failed to provide scientific reasons to justify the choice of a GHG reduction target of 20% by 2020, for instance by presenting long-term references scenarios to prove how the currently postponed reduction could nonetheless lead to achieving the overall 2° target (para. 52).

The Appellate Court judgment in Urgenda firmly reinforces the case for the employment of right claims in climate change litigation as linked to scientific evidence that mitigation action is vital for preventing dangerous climate change (Peel and Osofsky 2018).

3.2 Gloucester Resource Limited v. Minister for Planning

Australia is the world's largest exporter and fourth largest producer of coal. Yet, the country also holds a rather remarkable track record of successful attempts to halt fossil fuels projects, thus placing at the forefront of the permit climate litigation (McAllister 2008). *Gloucester Resource Ltd. v. Ministry of Planning* (also known as the *Rocky Hill* case) represents the first instance when an Australian court has denied the application for a coal mining project not only on the basis of unacceptable planning, visual and social impacts, but to prevent a new source of GHG emissions (Hughes 2019, p. 6).[9] In March 2010 a mining company (*Gloucester Resources Limited*, GRL) commenced extensive exploration, preparation and investigation into an 830 ha open-cut coal mine (the Rocky Hill Coal Project) in a valley near the small country town of Gloucester, New South Wales. Two years after the initial project planning, GRL applied to the Department of Planning and Environment (DPE) for development consent to begin operations. In 2016, the application was accepted, amendments were made, and the application was placed on public exhibition for three months. During the exhibition period, 2570 submissions were received, approximately 90% of which in opposition to the mine project. By the end of 2017, following DPE's recommendation, the amended application was rejected under Article 89 of the 1979 Environmental Planning and Assessment Act. The Planning and Assessment Commission, acting on behalf of the Minister for Planning, justified its refusal on three reasons, namely: (1) the incompatibility with existing, approved and preferred land uses in the selected area; (2) the visual impact of the project; (3) the overall lack of public interest.

GRL filed an appeal before the NSW Land and Environment Court, in person of Chief Justice Hon Preston. The local community action group, Groundswell Gloucester Inc. (GG), joined the Minister for Planning as respondents. On February 8, 2019, Chief Justice Preston dismissed the appeal and rejected the development application.

At the hearing, GG grounded the refusal request on the fundamental argument that GHG emissions generated by mine site activities, as cumulated with the

[9]Gloucester Resources Limited v. Minister for Planning [2019] NSWLEC 7.

burning of extracted coal, would be utterly "inconsistent with existing carbon budget and policy intentions to keep global temperature increases to below 1.5° to 2 °C above pre-industrial levels and would have a cumulative effect on climate change in the long term" (para. 422).[10] Such arguments have been grounded on IPCC's AR5 calculations on the global carbon budget. GG cited several carbon budget calculations to advocate that, given the relentless risk of overshooting the carbon budget in little more than two decades, "[m]ost of the world's existing fossil fuel reserves – coal, oil and gas – must be left in the ground, unburned", and "no new fossil fuel developments can therefore be allowed" (para. 447). More specifically, based on recent studies on global carbon budget's implications for fossil fuels sectors, to achieve the 50% probability of meeting the 2 °C target would require 62% of the global existing fossil fuel reserves to be not exploited, and 88% of coal reserves would not be burned at any rate.[11] Narrowed down at the regional scale, over 90% of Australia's existing coal reserves should not be burned in order for the country to operate consistently with the Paris 2 °C target—let alone the most stringent 1.5 °C target (McGlade and Ekins 2015). Thus, development of any new fossil fuel reserves, "no matter how small, is incompatible with any carbon budget assuming a 50% or better chance of the budget meeting the temperature target" (para. 449).

Importantly, CJ Preston firmly recognized the causal link between the Rocky Hill project's cumulative GHG emissions and global warming. The judge also accepted GG's evidence on the carbon budget approach required to achieve long-term temperature goals. Moreover, these arguments served as the basis for the *wrong time* test, according to which a new mining site such as the Rocky Hill project would "increase global total concentrations of GHGs at a time when what is now urgently needed, in order to meet generally agreed climate targets, is a rapid and deep decrease in GHG emissions" (para. 699). Looking at the project from a carbon budget perspective, CJ Preston quashed GRL's arguments that GHG emissions from the mine would represent only a fraction of total global emissions, which altogether contribute to the carbon budget, and therefore to climate change.[12] In the judge's view, multiple local actions are needed to adequately mitigate emissions at sources

[10] GG's arguments fundamentally relied on the expert testimony of Emeritus Professor Will Steffen, an Earth System scientist at the Australian National University, Senior Fellow of the Stockholm Resilience Centre and Member of the Climate Council of Australia. Steffen's testimony drew on both global and Australian publications, including reports from the Intergovernmental Panel on Climate Change, the Australian Bureau of Meteorology and the Commonwealth Scientific and Industrial Research Organisation (CSIRO), Australia's premier scientific research organisation.

[11] The global 2011–2050 carbon budget estimation as presented in Court equalled 300 Gt C. The presented study also showed that about 780 Gt C would be emitted as CO_2 were all of the world's existing fossil fuel reserves burned—about 2.5 times greater than the allowable budget. Importantly, GG inferred from this data that to achieve the Paris Agreement targets implies not only currently operating mines and gas wells to be closed before their economic lifetime, but also that no approved and proposed fossil fuel projects, based on existing reserves, shall be implemented.

[12] Relevant, CJ Preston extensively relies on the doctrine in Urgenda v. The Netherlands District Court's decision to reject this argument (paras. 521–524).

and remove GHGs by sinks. Hence regardless of its eventual economic benefits, the Rocky Hill project would not prove conducive to the rapid and deep GHG emissions reduction needed for Australia to adequately contribute to the overarching goals agreed upon under the Paris Agreement.[13]

3.3 Lliuya v. RWE

The claimant, Mr. Saùl Luciano Lliuya, is a resident in the city of Huaraz, situated right below lake Palcacocha, in the Peruvian Andes. Lliuya asserts that, despite previous measures, because of melt water from the overlying Palcaraju glacier and precipitation accumulate in the lake the water level in lake Palcacocha is now tipping a dangerous status, and therefore poses serious and imminent flood risks. In all likelihood, a flood wave from the lake would severely impair Mr. Lliuya's property. The defendant (RWE) is a German energy company, and is by far the largest GHG emitter in Europe. According to the Carbon Majors Report, RWE operations contributed to a degree of 0.47% from 1751 to 2010 to worldwide GHG emissions (Heede et al. 2014, p. 27). Thus, Mr. Lliuya is seeking compensation from RWE in proportion to the latter's actual contribution to the change of the state of aggregation of the glacial ice over lake Palcachoca.

Relevant, the claimant extensively quoted IPCC's AR5 observations to demonstrate the actual specific impacts of climate change on the Peruvian Andes Region as to temperature increase and tropical glaciers meltdown—demonstrated in AR5 with "very high confidence". In addition, evidence provided by the Peruvian government was displayed, reporting a 40% melting rate of Peruvian glaciers over the last 40 years, the melt water resulted thereby contributing to the formation of approximately 1000 new lakes. Thus, in the plaintiff's view the causal link between increase in GHG emissions, global warming and the increased risk of a detrimental *Glacial Lake Outburst Flood* (GLOF) in the area of lake Palcacocha has been scientifically established.[14] Given the causal link between a dangerous GLOF risk and global GHG emissions surge, the defendant should be held liable for any damages occurring to Mr. Lliuya's property in the proportional share of its historical GHG gas emissions, as mapped and quantified in the Climate Majors study.

[13] Furthermore, CJ Preston noted that a particular fossil fuel development may itself be a sufficiently large source of GHG emissions such that refusal of the development "could be seen to make a meaningful contribution to remaining within the carbon budget and achieving the long term temperature goal" (para. 554).

[14] In particular, the claimant refers to the Inventario de Glaciares del Perú, Ministerio de Agricultura y Riego Autoridad Nacional del Agua, Unidad de Glaciología y Recursos Hídricos, Huaraz, July 2014, p. 23, point 7.1.3.1, according to which "Comparing the glacier surface area measured in the national inventory of the decade of the 1970s and the results of the current inventory, the Cordillera Blanca has lost approximately 27% (195,75 km^2) of its total glacier area."

From a legal standpoint, such claim falls within the category of cumulative damages under German law, or the so-called cumulative causation, which relies on the fundamental "but-for test" (*conditio sine qua non*). According to this test, a specific event is to be viewed as a cause had the damage not occurred without it. In the cumulative causation theory, the act of an offender would be deemed as a cause if it would result in the damage in combination with actions of other offenders (Koch et al. 2011, p. 399). Relevant, the enrichment of the atmosphere with GHG emissions caused by RWE would constitute an active external influence causing the impairment of the claimant's estate under §1004 BGB. Thus, RWE shall alternatively:

(a) Bear the costs for adequate preventative measures to protect the property of the claimant against a glacial flood from lake Palcacocha, proportionally to its contribution to the damage (i.e., its share of global GHG emissions);
(b) Take adequate measures to ensure that the water volume of lake Palcacocha is reduced to an extent proportional to the defendant's contribution to the damage;
(c) Be ordered to pay an amount (17,000 euros), equal to the 0.47% of the estimated costs to drain lake Palcacocha as its contribution toward preventative measures adequate for the protection of the claimant's property.

The District Court of Essen rejected the claim, stressing the disputed existence of a linear causation chain between a particular source of GHG emissions, "the state of the climate as it presents itself today [...] and to one particular damage".[15] In the Court's view, it was not clearly demonstrated that "anthropogenic climate change and therefore the supposed flood risk of the glacial lake would not occur if the defendant's particular emissions were not to exist". Relevant, the Court utterly failed to address the main argument of the claimant, namely that the flood risk would be lower had RWE's emissions not existed (Kahl and Daebel 2019).

The claimant challenged the decision before the Regional Court of Hamm, which allowed the claim to proceed to the evidentiary stage.[16] The Regional Court was satisfied that the scientific studies presented by Mr. Lliuya were sufficient to justify further inquiries to assess whether RWE operations would constitute an active contributory cause to the GLOF risk in lake Palcacocha. Thus, the Regional Court mandated experts opinions to clarify: (1) *if* RWE activities have led to a higher GHG concentration in the atmosphere; (2) *if* such higher concentration has caused an increase in global temperature; (3) *if* such increase in global temperature has in turn accelerated the melting of the Palcaraju glacier so that the water volume in lake Palcacocha can no longer be contained by the natural moraine; (4) *if* RWE's share in the whole causation chain is measurable and calculable, and accounts for 0.47% of the total.[17] While the evidentiary phase is still pending, *Lliuya v. RWE* represents

[15] District Court Essen, Case No. 2 O 285/15 (December 16, 2016).

[16] Regional Court of Hamm, Case No. 2 O 285/15 (November 30, 2017).

[17] As no agreement could be reached between the parties as to the nominee joint experts, the Court appointed its own experts on September 2018, who will inspect the premises in Peru subject to the lawsuit. In particular, the claimants contended that the RWE's suggested experts views were largely not based on IPCC findings.

the first instance when scientific evidence on climate change will be openly discussed in Court as part of the evidentiary activities to establish causation for climate-related damages.

4 Conclusion: Is Climate Science Turning the Tide?

The ultimate aim of climate change litigation can be understood as the effort to translate the remarkable degree of scientific consensus on climate change into legal principles and political action (Peel and Osofsky 2015, p. 9). After beginnings characterized by abrupt failures, climate lawsuits are now gaining a more successful path by relying on different legal strategies, which are commonly grass-rooted in the scientific evidence of the risks caused by anthropogenic climate change (Nachmany and Setzer 2018). Within this backdrop, to demonstrate a clear-cut association between the mention of climate science and more favourable outcome for climate litigants still constitutes a Gordian knot, which requires further enquiry (McKormick et al. 2018). Nonetheless, as the cases analysed above display, in absence of specific provisions specifically enabling climate-related lawsuits, relevant steps forward have been taken towards embracing IPCC's, as well as other climate science findings as foundational to: (1) evaluate States conducts with regard to their climate long-term policies; (2) inform public authorities decisions with regard to specific, highly GHG intensive projects; (3) establish a liability pattern for climate-related damages. Where policy-makers tend to refute or neglect climate science as a key tool to develop policies in different fields, adjudicatory bodies are arguably becoming accustomed to climate science studies and findings, and to appraise them within the context of the (indeed frequent) references included in relevant legal texts.[18] Whether it would be appropriate for Courts to lead the way through the urgently needed widespread debate about global growth cuts and modifications of our societal way of life, as opposed to democratically legitimized institutions, however, remains an open question (Wegener 2019).

References

Allen M (2011) The scientific basis for climate change liability. In: Lord R et al (eds) Climate change liability: transnational law and practice. Cambridge University Press, Cambridge, pp 8–22
Butt D (2017) Historical emissions: does ignorance matter? In: Meyer LH, Sanklecha P (eds) Climate justice and historical emissions. Cambridge University Press, Cambridge, pp 61–79

[18] The relevance of the precautionary principle in climate change law is beyond debate (De Sadeleer 2016). The UNFCCC requires States to promote scientific research and requires the COP to periodically examine, inter alia, the scientific and technological knowledge. The Paris Agreement requires Parties to undertake rapid reductions in accordance with best available science (Hanekamp and Bergkamp 2016). The Treaty on the Functioning of the EU requires the EU to take the available scientific data into account when preparing environmental policy (art. 191).

Carbon Tracker Initiative (2018) Mind the gap: the 1.6 trillion energy transition risk
Collins M et al (2013) Long-term climate change: projections, commitments and irreversibility. In: Climate Change 2013: The Physical Science Basis. Contribution of Working Group I to the Fifth Assessment Report of the Intergovernmental Panel on Climate Change. Cambridge University Press, Cambridge, United Kingdom and New York, NY, USA, pp 1029–1136
De Sadeleer N (2016) The precautionary principle and climate change. In: Farber D, Peeters M (eds) Climate change law. Edward Elgar, Cheltenham, pp 21–31
Duffy M (2009) Climate change causation: harmonizing tort law and scientific probability. Temple J Sci Technol Environ Law 28(2):185–240
French D, Pontin B (2016) The science of climate change: a legal perspective on the IPCC. In: Farber D, Peeters M (eds) Climate change law. Edward Elgar, Cheltenham, pp 9–20
Ghaleigh NS (2016) Science and climate change law – the role of IPCC in international decision making. In: Carlarne CP, Gray KR, Tarasofsky R (eds) The Oxford handbook of international climate change law. Oxford University Press, New York, pp 56–71
Hanekamp JC, Bergkamp L (2016) The "Best Available Science" and the Paris agreement on climate change. Eur J Risk Regul 7(1):42–48
Heede R (2013) Tracing anthropogenic carbon dioxide and methane emissions to fossil fuel and cement producers, 1854–2010. Clim Change 122(1–2):229–241
Heede R et al (2014) Carbon Majors: Accounting for carbon and methane emissions 1854–2010 – Methods and Results Report
Hoegh-Guldberg O et al (2018) Impacts of 1.5°C global warming on natural and human systems. In: IPCC global warming of 1.5°C. World Meteorological Organization, Geneva
Hughes L (2019) The Rocky Hill decision: a watershed for climate change action? J Energy Nat Resour Law 37(3):341–351
Kahl W, Daebel MC (2019) Climate change litigation in Germany: an overview of politics, legislation and especially jurisdiction regarding climate protection and climate damages. Eur Energy Environ Law Rev 28(2):67–76
Koch HJ, Lührs M, Verheyen R (2011) Germany. In: Lord R et al (eds) Climate change liability: transnational law and practice. Cambridge University Press, Cambridge, pp 376–416
Le Queré C et al (2018) Global carbon budget 2018. Earth Syst Sci Data 10:2141–2194. https://doi.org/10.5194/essd-10-2141-2018
Markell D, Ruhl JB (2012) An empirical assessment of climate change in the courts: a new jurisprudence or business as usual? Florida Law Rev 64(1):17–86
Mastrandrea M, Schneider SH (2014) Climate change science overview. In Schneider SH et al (eds) Climate change science and policy. Island Press, pp 14–27
McAllister LK (2008) Litigating climate change at the coal mine. In: Burns W, Osofsky HM (eds) Adjudicating climate change: sub-national, national, and supra-national approaches. Cambridge University Press, Cambridge, pp 48–71
McGlade C, Ekins P (2015) The geographical distribution of fossil fuels unused when limiting global warming to 2°C. Nature 517:187–190
McKormick S et al (2018) Strategies in and outcomes of climate change litigation in the United States. Nat Clim Change 8:829–833
Nachmany M, Setzer J (2018) Global trends in climate change legislation and litigation: 2018 snapshot, Grantham Research Institute Policy brief – May 2018.
Peel J, Osofsky HM (2015) Climate change litigation: regulatory pathways to cleaner energy. Cambridge University Press, Cambridge
Peel J, Osofsky HM (2018) A rights turn in climate change litigation? Transnl Environ Law 7(1):37–67
Rogelj J et al (2018) Mitigation pathways compatible with 1.5°C in the context of sustainable development. In: IPCC global warming of 1.5°C. World Meteorological Organization, Geneva
Setzer J, Vanhala L (2019) Climate change litigation: a review of research on courts and litigants in climate governance. WIREs Clim Change 10:1–19
UNEP (2017) The status of climate change litigation – a global review

UNEP (2018) The Emissions Gap Report 2018. United Nations Environment Programme, Nairobi
Victor DG et al (2014) Introductory chapter. In: Climate change 2014: mitigation of climate change. In: Contribution of Working Group III to the Fifth Assessment Report of the Intergovernmental Panel on Climate Change. Cambridge University Press, Cambridge, United Kingdom and New York, NY, USA
Wegener BW (2019) Urgenda – world rescue by court order? The "Climate justice" movement tests the limit of legal protection. J Eur Environ Plann Law 16(2):125–147

‘Thieves, Shady Deals and Murder’: Water Theft, Buy-Backs and Fish Kills in the Murray Darling Basin of Australia

Janice Gray

Abstract Effective water management in the Murray Darling Basin of Australia has proved challenging. Water theft, problematic water buy-backs by the Commonwealth and fish kills, particularly those in the Menindee Lakes, have exposed weaknesses in the implementation and application of key legislation and governance instruments such as the Water Act 2008 (Cth) and the Murray Darling Basin Plan. Meanwhile, Basin health has continued to suffer. A more pro-active, multi-pronged approach to governance, particularly in relation to monitoring, compliance and enforcement would be advantageous as would greater transparency in the application of science-led or evidence-led research. Improvements in these areas are likely to better serve ecological integrity objectives.

1 Introduction

The *Water Act 2007* (Cth) (*WA*) and the Murray-Darling Basin Plan 2012 (MDBP) were introduced with high hopes. It was intended that the Act and Plan would significantly improve water management in the Australian Murray-Darling Basin (MDB) but achievements under these instruments have not lived up to expectation. Logistical, implementational and practical weaknesses have dogged their effectiveness.

This chapter focuses on three issues: (1) water theft, (2) water buy-backs and (3) fish kills to expose some of the weaknesses that have emerged in MDB water law and governance. It also reveals how those weaknesses are undermining ‘ecological integrity’ (Bosselmann 2008) and are instead supporting the continued over-exploitation of finite water resources, ultimately contributing to, ‘disintegrity’.

J. Gray (✉)
Faculty of Law, University of New South Wales, Sydney, NSW, Australia
e-mail: j.gray@unsw.edu.au; https://orcid.org/0000-0003-4129-1011

L. Westra et al. (eds.), *Ecological Integrity in Science and Law*,
https://doi.org/10.1007/978-3-030-46259-8_4

Indeed approximately 12 years on from the *WA*'s introduction and seven years on from the MDBP's, there is little or no evidence that the river system is much healthier (Jackson 2016; Grafton et al. 2018a, b). Leading scientists have claimed that the MDBP, whose aim is to 'return to an environmentally sustainable level of [water] extraction') (MDBP 2012 ch 5, s5.05), 'is failing to restore the river's health' (Davies et al. 2019). Rivers are running dry (Davies et al. 2019) while smaller scale water-dependent farmers and agriculturalists claim to be priced out of the water market (Hunt 2019), threatening their continued operation, and impacting on the nature and extent of food grown and produced (Gribbin 2019). Meanwhile, large, often multi-national investors, are entering the field and expanding the production of water-thirsty crops such as cotton and nuts (ABC Four Corners 2017; Aither 2019).

Although the terms, 'thieves, shady deals and murder' are more commonly associated with crime gangs and underworld characters, it is perhaps not too far off the mark to see the emerging MDB water management narrative in a similar, unfavourable (if somewhat colourful) way. Water has been 'stolen' by some irrigators. In other cases, it has been bought back for the environment under questionable commercial deals, and fish in the Menindee Lakes have been 'murdered' by their thousands. The picture is a disturbing one.

The chapter argues that there is an urgent need for a more pro-active approach to monitoring, compliance and enforcement and it calls for: (a) acknowledgement of the 'rationalist fallacy' (Brown et al. 2016) in water law and governance; (b) a multi-pronged approach to water reform and governance and (c) greater transparency in relation to science-led water law and governance.

2 Background

The MDB is Australia's food bowl (Gray 2019). It is simultaneously a recreational playground, the habitat of flora and fauna, and a culturally important place (MDBA 2019a, b). The river system is also significant in its own right (The Green Institute 2019; International Union for Conservation of Nature 2010). However, or perhaps because of this, the MDB has proved difficult to govern (Gray et al. 2017, ch 1; Carmody 2017).

The MDBP is a legislatively mandated tool of the *WA* and is designed to restore MDB health by: (a) setting a Sustainable Diversion Limit (SDL) restricting how much water may be taken from the basin; (b) facilitating the buying back of water entitlements for the environment; and (c) funding infrastructure/efficiency programs, so less water is used. The MDBP is in many ways, an *aspirationally* laudable instrument because its objectives are hopeful, progressive and 'green friendly' but its implementation thus far, has proved little short of catastrophic (Walker 2019, p. 5, p. 11).

Both the *WA* and the *MDBP* (Commonwealth instruments) operate in tandem with state-based laws, such as the *Water Management Act 2000* NSW (*WMA*) which sets up a comprehensive system of (a) water licensing through water entitlements and allocations and (b) water plans (Gardner et al. 2018, chs 2, 12, 15 and 25). With very few exceptions (see *WMA 2000* s52; *Water Management (General) Regulation* 2018) either an entitlement or allocation under a Water Access Licence (WAL) (see *WMA 2000* s 56) is required for lawful water abstraction in New South Wales (NSW). WALs, entitlements and allocations may be purchased on the market.

3 'Thieves': Water Theft

In 2017, the current affairs program, *Four Corners* aired allegations of possible water theft (ABC 2017). Some irrigators were thought to have taken water to which they were not entitled under their WALs and associated governance instruments. Public reaction to the allegations was swift (Fitzpatrick and Calvert 2017). Even if people were not greatly concerned about the environment, they did not like a thief.

In response to the theft allegations and growing public concern, the NSW Government commissioned the Matthews' Report. Matthews found NSW's water compliance and enforcement were 'ineffectual and require[d] significant and urgent improvement' (NSW Department of Industry 2017a, p. 4). In particular, he found that metering, monitoring and measurements of water extraction in the Barwon-Darling system were below standard (NSW Department of Industry 2017a, p. 4). Accordingly, water could be easily stolen because there were inadequate records of abstractions being kept. Matthews, therefore, recommended a 'no metering, no pumping' rule (NSW Department of Industry 2017a, p. 4) and suggested a 'systemic fix' (NSW Department of Industry 2017a, p. 4) that included setting up a separate Natural Resources Access Regulator (NSW Department of Industry 2017a). He also criticised the tardiness of the state regulator's responses (NSW Department of Industry 2017a, p. 4) and noted that 'the [water] industry's "social licence to irrigate" [was] at stake' (NSW Department of Industry 2017a, p. 5), so acknowledging a social and ethical dimension (not just a legal one) to the issue of water sharing.

The NSW Parliament's legislative response to the situation came in the form of the *Water Management Amendment Act 2018 (No 31)* which increased the maximum penalty for water theft to five million dollars for a company and $500,000 for an individual but did not introduce a total ban on unmetered water abstraction (NSW Department of Industry 2017b). Hence, floodplain water may still be harvested in NSW without monitoring. Levees and canals may 'legally' divert flood waters into irrigation storage tanks and away from rivers, thus contributing to ongoing environmental harm and ill effects for downstream users.

3.1 Barlow's Case

Barlow's case represents one of the few water theft prosecutions. Barlow ultimately pleaded guilty and in 2019, was sentenced (*WaterNSW v Barlow*) for taking water when an embargo imposed by *Temporary Water Restrictions (Upper Darling Basin)*

Order (No 2) 2014 under s 324 of *WMA* was in place. His conduct constituted an offence under *WMA* s 336C(1) and was, in common parlance, a form of water theft. Further, he took water when two of his pumps had broken meters; an offence under *WMA* s 91L(2). The penalty handed down was $189.491.00 plus costs.

In deciding the sentence Preston J took into account the objective seriousness of the offence and the subjective circumstances of the offender. In considering the former, he ultimately turned to the objects of the *WMA*. Those objects (that embed the concept of ecological sustainability in the statute) are comprised of principles (including the precautionary principle) (*Water Management Act 2000* s 3). One way that the *WMA* seeks to ensure those principles are fulfilled is by regulating water abstractions. Barlow's conduct ignored the relevant regulation and underpinning principles. In doing so, it contributed to the erosion of ecological integrity.

Whilst beefed-up criminal sanctions may deter actors such as Barlow from future re-offending, it may also be helpful to understand *why* Barlow breached in the first place. He may for example, have resented 'a legislated national program of reduced water for irrigation' (Walker 2019, p. 20) grievously threatening his future enterprises; an issue that water governance may need to better address.

4 'Shady Deals': Water Buy-Backs

Water buy-backs represent one mode by which the Commonwealth can acquire water for environmental watering. Once acquired by the relevant government department, the water is held and managed by the Commonwealth Environmental Water Holder (CEWH) (Slattery and Campbell 2018, p. 1).

Several buy-backs have proved problematic. Criticism ranges from the high prices paid for some licences, the types of licences/rights purchased (some yield little water), the below-par standards of due diligence and procurement procedures and the potential conflict of interest between vendors and purchasers, to a fear that multi/trans-national purchasers may not have the best interests of Australian society at heart (Hannam 2019; Slattery and Campbell 2018, p. 1).

4.1 Kia-Ora/Clyde

The Kia-Ora/Clyde buy-back has proved particularly contentious. Eastern Australia Agriculture Ltd (EAA), a large water holder with multinational investors (Bloomberg 2019), offered to sell the Department of Agriculture (DoA) its Overland Flow (OF) licences at its Kia-Ora and Clyde properties (Queensland). Its original asking price for the licences and storage was $27,880,000 (EAA Letter cited in Slattery and Campbell 2018, p. 8) but DoA ultimately purchased fewer megalitres of licences without storage for the higher price of $78,891,300 (Slattery and Campbell 2018, p. 8). DoA's final per megalitre purchase price was 25% higher than EAA's original, offered selling price.

Being floodwater, OF water is inherently unreliable (*Water Act 2000* (Qld) sch 4; Bureau of Meteorology 2016). It can, in the relevant part of Queensland, be diverted and stored (*Condamine-Balonne Water Resource Plan* 2019) but harvesting may be limited by a moratorium, water resource plan (WRP) or a wild rivers declaration to prevent harm to ecosystems and communities relying on such water (Bureau of Meteorology 2016). As a result of DoA's $78 million purchase, the CEWH may now harvest OF water and return it to the environment but if the OF water had not been purchased by DoA and had not been harvested, it would have, in all likelihood, been returned to the environment—for free.

Further, if OF licences are non-tradeable amongst irrigators, their value should be lower than many other types of licences. Yet, the price paid by DoA for OF licences represented 74% of the value of Kia-ora and Clyde properties (EAA Annual Report for the Year Ended 30 June 2017 cited in Slattery and Campbell 2018, p. 8). That neither storage nor land was included in the purchase price has led some commentators to question the wisdom of the purchase (Davies 2018)—a large outlay for a rather limited gain. What motivated such seemingly profligate expenditure, particularly in light of the CEWH's confirmation that the $79 million buy-backs 'have returned next to no water to the environment since they were purchased two years ago' (Middleton 2019a), has not been established.

This buy-back also raises potential conflicts of interest. A former Energy Minister was a co-founder and director of the vendor's parent company, Eastern Australian Irrigation (Davies 2019). He has denied receiving any benefit from the $78 million EAA buy-back; a buy-back that proceeded without an open tender and arguably tested the boundaries of procurement procedures (*Public Governance, Performance and Accountability Act 2013* (Cth) and *Commonwealth Procurement Rules* (2017)). Given that the CEWH allegedly provided general advice to DoA pointing out that that 'supplementary' (including OF) water was the least attractive purchase option, (Hasham 2019) one is led to ask, was such a purchase justifiable? It is perhaps timely that the Northern Basin Commissioner has been examining connections between political donations and buy-backs. He has made no findings but has observed that where conflicts of interest exist 'and they are not addressed, [conditions are] ripe for corruption' (Keelty quoted in Middleton 2019a).

4.2 Tandou

The buy-back of Webster's WALs for its Tandou property, south of Menindee (NSW), at a price of $A80 million—almost double the price that the Commonwealth's own research agency considered the licences were worth (Davies 2017)—is another contentious purchase. The Commonwealth's purchase, vigorously encouraged by the NSW government, involved no tender process, no cabinet approval and seemingly no advice (either sought or relied on) from other departments or agencies such as CEWH (Davies 2017). Whilst the Commonwealth's procurement rules make it possible to avoid going to tender in cases where 'exceptionally advantageous conditions … arise only in the short term', that position has not yet been firmly established in the Tandou case (*Procurement Rules 2017* (Cth) r 10.3(c)).

The Tandou buy-back arguably supports the Menindee Lakes Drainage Project in NSW and allows that state to reach its water saving target for environmental water by shrinking the lakes and reducing evaporation. Yet, re-vitalising the ailing MDB by draining the lakes is misguided (ASS 2019, p. 36). Ecological integrity is not simply about saving water. It is about restoring ecology.

In conclusion, none of the extravagant buy-backs of low reliability water licences have, thus far, assisted society to avoid crossing those thresholds that 'endanger… the basic integrity of the system' (Bruntland Report 1987, para 23).

5 'Murder': Fish Kills

In late 2018 and early 2019, there were three major fish kill events in the Menindee Lakes. Murray Cod (up to 70 years old) Golden Perch, Silver Perch and Bony Herring all died as part of the estimated million or more fish deaths.

Following the kills, two major reports were commissioned. The Vertessy Report concluded that the immediate cause of the fish kills was stratification of the water column in the weir near Menindee (Vertessy et al. 2019), leading to low levels of dissolved oxygen that denied the fish the oxygen they needed to survive (NSW Department of Industry 2019).

The Moritz Report agreed but also highlighted long and medium-term causes, blaming drought and excessive upstream diversions of water for irrigation (AAS 2019 Pt 4,7). It found that prior water releases from the Menindee Lakes contributed to the lack of local reserves, observing that it is important to maintain even low flows to help fish breed and to supply water to towns (AAS 2019, p. 23, 36).

5.1 Fish Kills and Excessive Water Abstractions

The Moritz Report concluded that the decline in long term flows is related to 'pumping environmental water, increased flood plain harvesting, policy changes in NSW in relation to the Barwon-Darling [WSP] and access to low flows and theft' (ASS 2019, p. 36). The result is that the 'maximum dry periods between low flow events have doubled' and may be up to 10 times longer while low flows themselves have been reduced by 70% (ASS 2019, p. 26), suggesting a link between governance and environmental harm.

Such a link was arguably affirmed by the Natural Resources Commission report on the Barwon-Darling's WSP which observed that the Barwon-Darling was 'an ecosystem in crisis' (NRC 2019, p. 1). Fish kills represent only one, very visual example of that ecological crisis. Such crises will continue if the volume of water abstracted by irrigators is not reduced. Those abstractions put pressure on the whole MDB system. Accordingly, WSP reviews need to address excessive water abstractions if ecological integrity is to be respected and crises such as fish kills, are to be averted.

5.2 *The SDL*

Beyond WSPs, another means of reducing extractions is to adjust the SDL. The SDL is 'the ultimate quantitative control imposed by the *WA*' (Walker 2019, p. 19) and is 'is intended to cap the volume of Basin water taken for consumptive use, such as for irrigated agriculture' so as to reflect 'an environmentally sustainable level of take' (ESLT) (Walker 2019, p. 19). According to the South Australian government and others, a level of take equivalent to a water recovery of 2750 gigalitres is inadequate to properly support key environmental values (Walker 2019, p. 383). That figure was the result of political compromise, not legitimate scientific evidence. A level of 3200 gigalitres had been previously mooted (also probably inadequate) but after irrigator protest, the MDBA revised the figure down. Some former MDA officials even told the South Australian Royal Commission they were instructed that the number 'had to have a 'two at the front' (Middleton 2019b).

Further, the scientific evidence that was used was not the best available. The SDL projections did not incorporate climate change. Consequently, the SDL was arrived at in an unlawful manner (Walker 2019, p. 247; *WA*). The SDL needs to be adjusted on the basis of best science, which would result in reduced water-take.

6 Conclusion

6.1 *Legal and Non-Legal Tools*

It is difficult to uphold ecological integrity if finite resources, such as water are used in a profligate manner. Hence, most countries have implemented systems of water sharing or allocation. In Australia, the system involves the licensing of water access. That system is challenged when some parties 'steal' water and take water outside the regulatory system.

Water theft, by definition, involves a party taking water to which he/she is not entitled. As a consequence, other actors (including the environment) will be unable to benefit from that same water. Yet, theft is likely to continue if parties believe that they will not be caught and if they are unconstrained by moral or ethical considerations. Effective monitoring (especially via metering), compliance and enforcement through the use of civil and criminal penalties (see Castilla-Rho et al. 2019, relying on Gelcich et al. 2006, Jones and Andriamarovololona 2008, and Ostrom and Walker 2005), is therefore important, while significant public condemnation and disapproval are also likely to act as a deterrent.

6.2 The Rationalist Fallacy

In the case of water theft, buy-backs, and fish kills, it will also be important to apply and enforce the law, including the law of procurement, civil and criminal fraud, restitution and water management but approaches that are too narrowly legalistic are unlikely to bring about the seismic changes that are necessary to protect water resources. It may, therefore, be helpful to go beyond law to consider why populist, right wing groups eschew environmentalism and to explore the *feelings* of disenfranchisement, disaffection and alienation that have been linked to the rejection of better environmental governance. Connecting with and understanding such feelings may prove vital because people do not necessarily act rationally. As Brown argues in the criminal justice context, developing rational policy models may not necessarily yield desired outcomes (Brown et al. 2016, p. 227). Rational policies and explanations are often simply not persuasive as Hochschild demonstrates with her real-life example of a person who lost his home to a sink-hole caused by a lightly regulated drilling company but nevertheless, still supported government deregulation and cuts to government spending on environmental protection (Hochschild 2012, p. 5).

6.3 Getting Into the Skin of Others

Consequently, sociological immersion research that 'gets into the skin' of people who reject policies and laws that would benefit them—the 'great paradox'—may be helpful (Hochschild 2012, pp. 8–20). It may assist appreciation of 'their "deep story", a narrative as *felt*' (Hochshild 2012, p. ix). From such an appreciation better, more inclusive, non-judgemental policies, law and governance may emerge. Hence, engaging with communities at a grass roots level, in order to understand *how* and *why* they can, on one hand, claim to care about their children's future but simultaneously continue to elect politicians whose parties consistently fail to address issues such as: climate change; the high number of water licences; a SDL that is set too low to achieve sustainable outcomes; and more generally, the importance of ecological integrity, may ultimately assist water governance.

Immersion research may reveal what is behind the conduct and views of people such as water thieves. Is it, for example, anger, social anxiety or resentment about their position in society that drives their conduct? Such research may also help answer the question of why people (many of whom are not well off themselves) continue to support governments that pay companies such as Webster and EAA unjustifiably large sums of money for buy-backs that have not improved river system health? On Hochschild's findings, the answer may lie not in rational explanations but rather in the sense of alienation and disaffection that such people *feel* more generally. She found that many people who rejected greater regulation and environmental policies that would help them, were people who felt set adrift by government and as victims (a term they were unlikely to use about themselves) felt resentful and unfairly done by (Hochschild 2012, p. 232). Paradoxically, they supported the very governments and companies that contributed to their feelings of being ignored, left out and disenfranchised.

6.4 Cutting in Line

The disenfranchised also saw groups such as environmentalists as cutting in line, getting special deals, unfairly reaping benefits and failing to respect the hard work of the alienated, conservative, petit bourgeoisie of which they were part; hard work that has traditionally been the tool of social assent. The implication is that, to the alienated conservative, a tree or river may be better off than they—prioritised ahead of them. Hence, hostility grows and those supporting causes such as the environment become the enemy of the disaffected (Hochschild 2012). In turn, the enemy's policies and laws are rejected. However, coming to a genuine understanding of how the alienated and disaffected *feel* (finding the 'empathy wall') (Hochschild 2012, pp. 5–8) may allow that group to be brought into the circle of collaborative decision-making, helping to build a broader coalition for change and better governance.

And it is a broad coalition that is needed. The left's views on water justice must meet the views of more conservative groups (and vice versa) as a matter of 'political strategy and program' (Brown et al. 2012, p. 245) if workable and effective outcomes supporting ecological integrity are to be achieved. Business, climate scientists, trade unions, economists and others need to come together as a force for change, although different actors may move in and out of the coalition at different times.

6.5 Re-Engaging Government

Further, re-engaging government, so that it demonstrates moral and ethical leadership and does not usurp its responsibility (by, for example, simply letting the market decide) will be critical (see Gray in Westra, Taylor and Michelot (eds) 2012; Gray 2013; Gray and Lee 2017; also Holley and Sinclair 2016). But how is government to be re-engaged, when many people feel that 'big government' is the problem in the first place, as the UK Brexit debate and Donald Trump's election revealed? One method is through the use of public protest and resistance, as Lock the Gate and the Knitting Nanas demonstrate (Gray 2014, 2016; Gunningham 2019). They have united disparate spheres of influence, acted as environmental defenders and brought pressure to bear on government in the water and fracking context for some years (Gray 2016). More recently, Extinction Rebellion has employed similar strategies with significant success, pressuring government (because resistance alone is not enough) and demonstrating how separate 'weak strands of influence' (Braithwaite and Drahos 2000, p. 33) may become stronger strands when joined together.

Pressure may also be brought to bear on government through serious, investigative journalism. In the UK, for example, the voting records of parliamentarians on important climate issues have linked voting patterns to affiliations with particular industries and the receipt of benefits (The Guardian 2019, pp. 18–21). Journalism of this kind may help catalyse change, particularly if it invigorates newly interested actors and provides protest movements with greater justification for their actions.

Oddly, two seemingly polar opposite approaches (engaging with those who reject government and simultaneously militating for a greater governmental role) may need to operate in tandem if better water outcomes are to be achieved in relation to theft, buy-backs and fish kills.

6.6 *Science-Led Law and Governance*

Finally, there also needs to be a more honest approach to science-led law and governance. To say, for example, that the SDL and ESLT were arrived at through science-led research is simply untrue (Walker 2019, p. 243). Science-led research should have been relied upon to comply with the *WA*. However, this is not to suggest that the best available science would have resulted in a flawless or objective MDBP. Scientific research is mediated in a political field. It may, and often is, subjective—it is not without values (Don Brown chapter one in this book) but that is not an argument for ignoring or expunging it. It simply means that subjectivity needs to be acknowledged in the hard sciences in the same way as it is conventionally acknowledged in the humanities and social sciences. Water theft, buy-backs and fish kills will be better governed if it is.

References

ABC Four Corners (2017) Pumped: Who's Benefitting from the Billions Spent on the Murray-Darling? https://www.abc.net.au/4corners/pumped/8727826. Accessed 15 Oct 2019

Aither DEWLP (2019) Water Supply and Demand in the Southern Murray Darling Basin Report. https://waterregister.vic.gov.au/images/documents/Water-Supply-and-Demand-Report_Aither_FINAL.pdf. Accessed 11 Nov 2019

Australian Academy of Science (AAS) (Moritz Report) (2019) Investigation of the Causes of Mass Fish Kills in the Menindee Region NSW over the Summer of 2018–2019. pp 23, 36. https://www.science.org.au/files/userfiles/support/reports-and-plans/2019/academy-science-report-mass-fish-kills-digital.pdf. Accessed 19 Aug 2019

Australian Academy of Science (ASS) (Moritz Report), Investigation of the Causes of Mass Fish Kills in the Menindee Region NSW over the Summer of 2018–2019, 18 February, 2019, p 2, Pt 4 p 17. Accessed 19 Aug 2019

Bloomberg (2019) Eastern Australia Agriculture Pty Ltd. https://www.bloomberg.com/profile/company/0367577D:AU. Accessed 3 Oct 2019

Bosselmann K (2008) The principle of sustainability. Ashgate, Farnham

Braithwaite J, Drahos P (2000) Global business regulation. CUP, Cambridge

Brown D et al (2016) Justice reinvestment: winding back imprisonment. Palgrave Macmillan

Brown D, Cuneen C, Schwartz M, Stubbs J, Young C (2012) Justice reinvestment: winding back imprisonment. Palgrave Macmillan, Basingstoke

Brundtland Report, Report of the World Commission on Environment and Development: Our Common Future. https://sustainabledevelopment.un.org/content/documents/5987our-common-future.pdf. Accessed 11 Nov 2019

Bureau of Meteorology (2016) Jurisdictions: Runoff, Harvesting. http://www.bom.gov.au/water/nwa/2016/mdb/media/Jurisdictions_runoff_harversting.pdf. Accessed 12 Sept 2019

Carmody E (2017) The unwinding of water reforms in the Murray darling basin: a cautionary tale. In: Holley C, Sinclair D (eds) Reforming water law and governance: from stagnation to innovation in Australia. Springer, Singapore, pp 35–56

Castilla-Rho JC, Rojas R, Andersen MS, Holley C, Mariethoz G (2019) Sustainable groundwater management: how long and what will it take? Glob Environ Chang 58:101972. Available at www.elsevier.com/locate/gloenvcha

Davies A (2017) $78m Buyback of Darling Water was Nearly Double its Valuation. The Guardian. https://www.theguardian.com/australia-news/2017/oct/26/78m-spent-on-darling-water-buy-back-nearly-double-its-valuation. Accessed 22 Sept 2019

Davies A (2018) Cotton company reaped $52m windfall in sale of water rights to government. The Guardian. https://www.theguardian.com/environment/2018/mar/02/cotton-company-reaped-52m-windfall-in-sale-of-water-rights-to-government. Accessed 3 Oct 2019

Davies A (2019) Questions over companies chosen for $200m of Murray-Darling water buybacks. The Guardian. https://www.theguardian.com/australia-news/2019/apr/17/questions-over-companies-chosen-for-200m-of-murray-darling-water-buybacks. Accessed 11 Nov 2019

Davies A, Bowers M, Ball A, Evershed N (2019) Murray darling: when the rivers run dry. The Guardian. https://www.theguardian.com/environment/ng-interactive/2018/apr/05/murray-darling-when-the-river-runs-dry. Accessed 2 Aug 2019

Fitzpatrick K, Calvert A (2017) Water theft allegations prompt government review of Murray-darling basin. Huffington Post. https://www.huffingtonpost.com.au/2017/07/30/water-theft-allegations-prompt-government-review-of-murray-darli_a_23056383/?guccounter=1&guce_referrer=aHR0cHM6Ly93d3cuZ29vZ2xlLmNvbS8&guce_referrer_sig=AQAAABk86WlXi2BcXfE_busSeCNslajD_2AaYID4IgbpJPwOE1KPpcas5xUUb6BpxxVfWQBd9NGPSk1YSg-bRvmG4DAgAeQaUtTTW. Accessed 15 Sept 2019

Gardner A et al (2018) Water resources law. LexisNexis, Sydney

Grafton Q et al (2018a) The Murray-darling plan is not delivering: there's no more time to waste. The Conversation. https://theconversation.com/the-murray-darling-basin-plan-is-not-delivering-theres-no-more-time-to-waste-91076. Accessed 2 Aug 2019

Grafton Q et al (2018b) Murray darling basin declaration. https://murraydeclaration.org/the-declaration. Accessed 2 Aug 2019

Gray J (2012) The legal framework for water trading in the Murray-darling basin: an overwhelming success? Environ Plann Law J 29(4):328–348

Gray J (2013) Dollars and dreams: legal aspirations and report cards in the Murray-darling basin of Australia. In: Westra L, Taylor P, Michelot A (eds) Confronting ecological and economic collapse: ecological integrity for law, policy and human rights. Routledge, Oxon

Gray J (2014) Frack Off! law, policy, social resistance, coal seam gas mining and the earth charter. In: Westra L, Vilela M (eds) The earth charter, ecological integrity and social movements. Routledge, Oxon

Gray J (2016) Trans-jurisdictional Water Governance in the context of unconventional gas mining: the Australian experience. In: Gray J, Holley C, Rayfuse R (eds) Trans-jurisdictional water law and governance. Routledge, Abingdon

Gray J (2019) Ecological integrity as an alternative frame for the water, unconventional gas and food Nexus. Jurimetrics 52(2):193–232

Gray J, Lee L (2017) Water entitlements as property: a work in progress or watertight now? In: Holley C, Sinclair D (eds) Reforming water law and governance: from stagnation to innovation in Australia. Springer, Singapore, pp 101–122

Gray J, Holley C, Rayfuse R (2017) The challenge of trans-jurisdictional water law and governance. In: Gray J, Holley C, Rayfuse R (eds) Trans-jurisdictional water law and governance. Earthscan, Routledge, pp 3–18

Gribbin C (2019) Shoppers warned to brace for higher food prices as Murray-darling basin faces increasing dry. ABC News. https://www.abc.net.au/news/2019-09-05/farmers-shoppers-warned-of-impact-of-murray-darling-basin-dry/11477182. Accessed 11 Nov 2019

Gunningham N (2019) Extinction Rebellion, Business and People Power, Oral Presentation, Kings College London, 25-9-2019

Hannam P (2019) Joyce's 'Goanna Water' deal raises queries after buyback doubles price. SMH. https://www.smh.com.au/environment/sustainability/questions-raised-about-joyce-s-goanna-water-buyback-deal-20180223-p4z1gw.html. Accessed 7 Oct 2019

Hasham N (2019) Water Buyback Criticism "a Joke", says Company at Centre of Scandal. SMH. https://www.smh.com.au/politics/federal/water-buyback-criticism-a-joke-says-company-at-centre-of-scandal-20190424-p51gu1.html. Accessed 22 Sept 2019

Hochschild A (2012) Strangers in their own land. New Press, New York

Holley C, Sinclair D (2016) Governing water markets: achievements, limitations and the need for regulatory reform. Environ Plann Law J 33(4):301–324

Hunt P (2019) Murray Darling Basin's shut down as dairy prices go sky high. The Weekly Times. https://www.weeklytimesnow.com.au/agribusiness/dairy/murray-darling-basins-dairy-shutdown-as-water-prices-go-sky-high/news-story/6a29d771b97ab7516cc1481453344f56. Accessed 2 Aug 2019

IUCN (2010) Draft Universal Declaration on the Rights of Mother Earth. https://www.iucn.org/content/draft-universal-declaration-rights-mother-earth. Accessed 9 July 2019

Jackson WJ (2016) Australia: State of the Environment. https://soe.environment.gov.au/sites/default/files/soe2016-overview-launch-version328feb17.pdf?v=1488792535. Accessed 7 Oct 2019

Middleton K (2019a) Keelty Warns River "Ripe for Corruption". The Saturday Paper. https://www.thesaturdaypaper.com.au/news/politics/2019/04/27/keelty-warns-river-ripe-corruption/15562872008055. Accessed 3 Oct 2019

Middleton K (2019b) Dead River, The Saturday Paper, 2-8 February, 2019

Murray-Darling Basin Authority (2019a) Discover the Basin. https://www.mdba.gov.au/discover-basin. Accessed 9 July 2019

Murray-Darling Basin Authority (2019b) Condamine-Balonne Water Resource Plan. https://www.mdba.gov.au/publications/mdba-reports/condamine-balonne-water-resource-plan. Accessed 3 Oct 2019

Murray Darling Basin Plan (2012) Australian Government, Federal Register of Legislation. Available at https://www.legislation.gov.au/Series/F2012L02240 Accessed 17 May 2020

Natural Resources Commission (NRC) (2019) Final Report Review of the Water Sharing Plan for the Barwon-Darling Unregulated and Alluvial Water Sources 2012 September 2019. https://www.nrc.nsw.gov.au//publications. Accessed 17 May 2020

NSW Department of Industry (2017a) Independent NSW Investigation into Water Management and Compliance. https://www.industry.nsw.gov.au/__data/assets/pdf_file/0016/120193/Matthews-interim-report-nsw-water.pdf. Accessed 12 Aug 2019

NSW Department of Industry (2017b) Securing Our Water: NSW Government Water Reform Action Plan. https://www.industry.nsw.gov.au/__data/assets/pdf_file/0016/136204/nsw-government-water-reform-action-plan.pdf. Accessed 27 Aug 2019

NSW Department of Primary Industries (2019) Fish Death Interim Investigation Report, January, 2019 https://www.dpi.nsw.gov.au/fishing/habitat/threats/fish-kills/Fish-death-interim-investigation-report.pdf. Accessed 12 Sept 2019

Slattery M, Campbell R (2018) That's not how you haggle. The Australia Institute. https://www.tai.org.au/sites/default/files/P502%20That%27s%20not%20how%20you%20haggle.pdf. Accessed 11 Sept 2019

The Green Institute (2019) Rebalancing Rights: Communities, Corporations and Nature. https://www.greeninstitute.org.au/wp-content/uploads/2019/03/Green-Institute-Publication-Rebalancing-Rights.pdf. Accessed 9 July 2019

The Guardian (2019) The Polluters Series, 12-11-2019. Partially available https://www.theguardian.com/environment/2019/oct/11/tory-mps-five-times-more-likely-to-vote-against-climate-action. Accessed 14 July 2019

Vertessy R, Barma D, Baumgartner L, Mitrovic S, Sheldon S, Bond N (2019) Independent Assessment of the Fish Deaths in the Lower Darling Final Report 29 March 2019. https://www.mdba.gov.au/sites/default/files/pubs/Final-Report-Independent-Panel-fish-deaths-lower%20Darling_4.pdf. Accessed 10 Sept 2019

Walker B (2019) Murray-Darling Basin Royal Commission Report. https://www.mdbrc.sa.gov.au/sites/default/files/murray-darling-basin-royal-commission-report.pdf?v=1548898371. Accessed 6 Sept 2019

Legislative Citations

Procurement Rules 2017 (Cth)
Public Governance, Performance and Accountability Act 2013 (Cth)
Temporary Water Restrictions (Upper Darling Basin) Order (No 2) *2014*
Water Act 2000 (Qld)
Water Act 2007 (Cth)
Water Management Act 2000 (NSW)
Water Management Amendment Act (No 31) 2018 (NSW)
Water Management (General) Regulation 2018 (NSW)

Case Citation

WaterNSW v Barlow [2019] NSWLEC 30

Refugees at Sea in a Hostile World

Joseph W. Dellapenna

Abstract In a world where hundreds of refugees are drowning at sea and thousands are being turned away at the border, the question arises not only as to why so many nations have become unwelcoming to refugees early in the twenty-first century, but as to whether national conduct violates international, regional, and national laws relating to refugees. To explore this question, this chapter not only describes the laws that are applicable to persons qualifying as refugees, but also seeks to set those laws in an historical context dating back millennia, with special emphasis on the experiences during and after World War II that gave birth to the laws now being flouted by so many nations. The chapter focuses on Europe, the crucible that largely gave rise to modern refugee law and one of the major locations of the current refusals to fulfil those laws. Special emphasis is given to Italy as one of the frontline states most deeply challenged by the ongoing refugee crisis.

On 25 July 2019, about 150 "migrants"[1] drowned in a shipwreck off the coast of Libya (Momigliano 2019); by that date, the toll of drowned "migrants" in the Mediterranean had already reached at least 423 for 2019. Weeks before, 44 "migrants" were killed in an airstrike on a detention camp in Libya (Walsh 2019). One could easily add many more accounts of dying "migrants" trying to reach Europe or other potential safe havens in recent decades. This chapter could focus on any of the "front-line states" in the European refugee crisis (Italy, Greece, and Spain). It focuses on Italy not because it is worse, but because it has become the focal point for such problems in 2019.

Migration and the flight of refugees is not a new story. Nor is resentment or worse directed at migrants. In order to understand events in Europe, this chapter begins with a brief overview of the history of refugees fleeing violence to

[1] Some, at least, of them might have qualified as refugees, and not merely as "migrants."

J. W. Dellapenna (✉)
Ardmore, Pennsylvania, United States

L. Westra et al. (eds.), *Ecological Integrity in Science and Law*,
https://doi.org/10.1007/978-3-030-46259-8_5

demonstrate that what has been happening currently is not unusual. There follows an analysis of the laws applicable to refugees today and the extent to which these laws have been followed. This is not worse than elsewhere, such as Australia (Cave 2019) and the United States (Ahmed and Villegas 2019; Kingsley 2019). Space does not allow discussion of these other situations.

1 Migrants and Refugees Before the Twentieth Century

Humans have been moving since the species began. *Homo sapiens* migrated across Africa and spread rapidly to all other habitable continents (*e.g.*, Harvati et al. 2019). People moved without conflict into lands not already occupied by humans or hominids. That possibility disappeared millennia ago. People moved, then and now, to seek better opportunities in life, or to flee problems such as famine, floods, or a changing climate.

Migration often involves violence. Sometimes the violence came from the migrants conquering or displacing prior inhabitants (Hermann 1954). An example is the Europeans who conquered the Western Hemisphere from the sixteenth to the nineteenth centuries. In other cases, the migrants were fleeing violence, violence that targeted individuals or that targeted ethnic or other groupings of people. Migrants fleeing violence have always been a special case, bearing the special name "refugee." Sometimes, when enough refugees settle together, they retain an ethnic cohesion even if they are not able to establish a new homeland. The best-known example is the Jewish *diaspora,* which began with the Babylonian exile of 586 BCE, and culminated in the Roman expulsion of the Jews from Palestine in the first and second centuries (Ehrlich 2018). Jews would not have a nation of their own again until 1948, yet despite ongoing persecution they survived as a distinct people. Or consider the people variously called "Gypsies," "les Bohèmes," "Roma," "Romani," etc. because of their supposed place of origin, who in fact are refugees from India (Fraser 1995).

The industrial revolution generated two phenomena that created unprecedented flows of refugees. First, it enabled vast population increases in the industrialized countries (Bar and Leukhina 2010). Vast numbers of these people voluntarily migrated from rural areas to cities, and also to other countries, seeking a better life (Hatton and Williamson 1994). Second, it brought on a series of violent social revolutions, each of which produced larger numbers of refugees. The first, the American Revolution (1775–1783) led to the flight of "Loyalists" to Canada. The French Revolution (1789–1815), produced what was seen at the time as numerous refugees (Carpenter and Mansel 1999). 1848 generated refugee flows across Europe, with many of the refugees eventually migrating to North or South America (*e.g.*, Tóth 2014).

2 Refugees in the Twentieth Century

The pattern of violent social revolutions generating refugee flows carried over into the twentieth century with the Mexican Revolution (1910–1920) (*e.g.*, Pace 1974), and the Russian Revolution (1917–1921) (*e.g.*, Hassell 1991). The twentieth century also saw the culmination of industrialized warfare. Industrialized warfare had arisen in the nineteenth century, generally without creating large refugee flows either because the wars were geographically isolated (Crimea, 1853–1856; Schleswig-Holstein, 1864; the Italo-Turkish War, 1911–1912), or because the wars were brief (the Austro-Sardinian War, 1859; the Seven Weeks War, 1866; the Franco-Prussian War, 1870).

The pattern of industrialized warfare changed in the mid-nineteenth century with the American Civil War (1861–1865). This war, in the heart of a growing industrial power, was not brief. Improved weapons made battles deadlier, yet improved transportation (the railroads) enabled both sides to replace their losses (Clark Jr. 2004). As a result, armies on both sides took to "digging in"—trench warfare. This pattern repeated itself in the Balkan Wars (1912–1913), and repeated itself with a vengeance World War I, perhaps the most stupidly fought war in history (1914–1918) (Bull 2014). By World War II (1937–1945), the militaries of the world (or at least some of them) had learned to break the stalemates of World War I, exploiting new technologies of mobility (tanks and mechanized warfare), of means of bringing force to bear (airplanes and self-propelled guns), and the oldest technique of all, surprise (shockingly neglected in World War I) (Clark 2016). The result in each of these wars were unprecedented numbers of military casualties and also vastly increased numbers of civilian deaths.

The industrialized warfare of the twentieth century extended to the systematic killing of civilians, sometimes only a few to make others flee, sometimes in an effort to eliminate an entire people. This was not something new. Many of the colonial enterprises had involved such "ethnic cleansing" (Hinton et al. 2014). But the scale and efficiency of the industrialized killing eventually led to the invention of a new word to describe the practice—genocide (Schabas 2000). Several notorious, large-scale genocides occurred during the period of the World Wars: the Armenian Genocide (1915–1918) (Kévorkian 2011); the Ukrainian Famine (1930–1933) (Conquest 1986); and the "Holocaust" (1940–1945) (Laqueur 2001).

Ethnic cleansing could happen without mass killings, as it did (more or less) at least twice: the Greco-Turkish population exchanges (1922–1923) (Clark 2009); and the Nazi-Soviet population exchanges (1939–1940) (Schechtman 1947). There was significant incidental killing, but in general they were well organized, successfully moving unwanted people to new locations where they were, if not welcomed, as least accepted. In the process, communities that had existed for hundreds (in the case of the Greeks, thousands) of years were obliterated. Nor did the concept of genocide capture all of the mass killings of the twentieth century: Stalin's purges (1934–1940) (Conquest 1968); the "Great Proletarian Cultural Revolution" (1966–1976) (Dikötter 2016); or Pol Pot's Cambodia (1974–1979) (Kiernan 2008), to name just the worst examples. For each of these, the mass killing were done for purposes other than the elimination of an ethnic, racial, or religious group (Genocide Convention 1948, art. II).

The wars and mass killing induced huge flows of refugees. World War II was particularly productive of refugees because the Germans moved vast numbers of people around Europe for their nefarious purposes. They moved people in order to kill them using industrialized (assembly line) processes (Kogon 1950). They held soldiers as long-term prisoners (many of whom died in captivity) (MacKensie 1994). And they sought to exploit the labor of millions of people, including people marked for death or captured soldiers, but also people enslaved solely for their labor (von Plato et al. 2010). The result at the end of the war was masses of what were euphemistically termed "displaced persons" (Wyman 1998).

Other countries than Germany (particularly the USSR) contributed to the displacing of people at the end of World War II (Pohl 1999), but the Germans were the largest contributor. There were at least 11,000,000 displaced persons in Germany alone at the end of the war (Antons 2014), and millions more scattered across Europe from the various German transports. Many from Central or Eastern Europe preferred becoming stateless refugees to returning to their prewar homes, both because those who survived in place had often collaborated in the displacement and because the refugees were unwilling to return to, or pass under, Stalin's rule (Bodor 1950).[2] The mass expulsion of ethnic Germans from Central and Eastern Europe added millions more to the stateless "floating" population (Douglas 2013).[3]

The upheavals of World War II were not the end of the wars and revolutions of the twentieth century, many of which generated large numbers of refugees. Until the twenty-first century, most of those conflicts produced flows of fleeing people who sought refuge in Europe. These were other, generally smaller conflicts that produced local refugee situations but did not produce a global crisis largely because such refugees often were able to find shelter in nearby countries that were culturally and ethnically similar. The size and nature of the displacement from World War II, however, called for and produced a global response, a response that largely remains in place today.

3 The UN Refugee System

Realizing the scale of population displacement during World War II, the western Allies set about to create an orderly and effective way of dealing with the refugees as early as 1943. Although the United Nations did not yet exist formally, the Americans and the British were using the term "United Nations" to denote their joint efforts, so when they undertook to create a refugee system in 1943, they named

[2] For example, in the heart of Olympiapark in Munich there is a small wooded enclosure where a former prisoner of war now known only as Timofey chose to live as a hermit for 27 years after the war rather than return to the USSR.

[3] While the Japanese did not transport so many people around Asia during their war, the same patterns did play out there, including the expulsion of millions of ethnic Japanese from Korea and Taiwan after the war.

it "the United Nations Relief and Rehabilitation Agency" (UNRRA) (UNRRA Agreement 1943; Fox 1950). When the UN was institutionalized in 1945, it took over the agency. UNRRA operated broadly across Europe and in China until 1948, working with displaced persons of every kind except ethnic Germans or ethnic Japanese, whatever their plight. Resettling and nurturing the latter groups became a problem solely for two nations already destitute after losing the war.

In 1946, member states of the UN drafted a treaty to create the International Refugee Organization (IRO Constitution 1946). After 26 states ratified it, the Organization began operating in 1948. As with UNRRA, the Organization operated extensively in Europe and eastern Asia, again excluding ethnic Germans and ethnic Japanese. Once the displaced persons from World War II were resettled, it ceased to operate in 1953. A broader responsibility for refugees across the world was passed onto the Office of the UN High Commissioner for Refugees, an office created by resolution of the General Assembly in 1950 (Statute of UNHCR 1950). UNHCR was, and is, authorized to provide international protection and assistance on a non-political, humanitarian basis to refugees everywhere, while seeking permanent solutions for them. Today it is responsible for about 70,000,000 people "in need of protection," 26,000,000 of whom it considers refugees in the formal sense (UNCHR Figures at a Glance 2019).

It was against this background that the member states of the United Nations drafted the *UN Convention on the Status of Refugees* (1951 Convention). It came into force in 1954 for Europe and eastern Asia (again excluding Germans and Japanese) and was set to expire when all persons who became refugees before 1 January 1951 were resettled. Today it has been ratified by 145 states (the ethnic, geographic, and temporal limits were dropped by a 1967 protocol). The US has ratified the protocol, but not the convention.

Under the convention, a refugee is defined as "a person with a well-founded fear of being persecuted for reasons of race, religion, nationality, membership of a particular social group, or political opinion," so long as the persons are outside country of their nationality and are unable to or, owing to such fear, are unwilling to accept the protection of that country or to return to it (1951 Convention, art. 1).[4] Excluded from this definition are economic or environmental refugees (in 1951, the possibility of environmental refugees had not yet been acknowledged).

The convention sets forth the rights and duties of refugees. The only duty for refugees is to obey the laws of receiving states (art. 2). In contrast, extensive rights are recognized for refugees. They have the right to:

1) Have their personal status (marriage, etc.) recognized (art. 12);
2) Free access to courts and to administrative assistance (arts. 16, 25);
3) Free movement and choice of residence within the country (art. 26);
4) Identity papers and travel documents (arts. 27, 28); and
5) The possibility of assimilating and naturalizing (art. 34).

[4] If the refugee is stateless (without any nationality), the relevant country is the refugee's former place of habitual resident.

In addition, receiving states are prohibited from:

1) Discriminating against refugees in general (art. 3), and specifically in:
    a. Rationing (art. 20);
    b. Education (art. 22);
    c. Public relief and assistance (art. 23);
    d. Labor (arts. 15, 17, 18, 19, 24); and
    e. Taxes (art. 29).
2) Taking exceptional measures based on a refugee's nationality (art. 8);
3) Punishing refugees for illegal entry if they present themselves "without delay" (art. 31); and
4) Expelling refugees or forcibly returning them to from whence they fled (arts. 32, 33) (*nonrefoulement*). (*See* Advisory Opinion 2007).

Receiving states may only limit refugee rights in the interests of national security (art. 9).

4 The European Architecture of Refuge

European leaders set about after World War II to create an institutional structure to ensure that Europe would not cause a third world war. What began as a small group of states cooperating on a narrow set of problems—the European Coal and Steel Community (1952) has now grown into a broad union that includes most nations on the continent and covers a wide range of responsibilities, the European Union. In the process, they created a zone of peace, prosperity, and stability adjacent to regions of instability, poverty, and strife. No wonder ever larger numbers of people from peripheral regions seek entry into the EU, or that one of the areas of responsibility for the EU as a whole is refugee policy.

Within the EU, national refugee policies pre-existed the EU and still persist within the limits set by EU policy. The trend of these national policies changed over time. Initially there was a (sometimes reluctant) willingness to resettle displace people left over from the war. While they might speak different languages and perhaps eat strange foods, they weren't so culturally different. Assimilation was relatively easy and the process went relatively smoothly (Murphy 1952). In the immediate aftermath of World War II, moreover, there was a shortage of adults, particularly of men, that refugees could help fill. Some countries, particularly Germany, welcomed foreign "guest workers" on the understanding that that they would only be temporarily in the country (Chin 2007). Guest workers were discouraged from bringing families and few, if any, efforts were made to assimilate them. But they did bring families. They and their families were different in culture, language, and religion. The end result was a distinct, imperfectly assimilated permanent minority population measuring in the millions.

The accession of eastern European countries to the EU facilitated a further migration of millions of workers into western Europe, creating additional tensions.

Many others sought entry into the EU, bringing in people very different in language, culture, and habits, and perhaps religion. Some seeking entry from the periphery were looking for economic opportunities; others were refugees fleeing wars, persecution, or natural disasters. There were wars enough, both within Europe and nearby, to generate refugee flows in the millions (McPherson 2019).

The EU gained responsibility over refugee policy when the *Treaty of Amsterdam* (entered into effect in 1999) gave the European Council of Ministers authority to develop a common refugee policy (Treaty of Amsterdam 1997, art. 73k). That policy has several elements, each arguably consistent with the 1951 *Convention* (which had been ratified by all EU member states). One feature could be termed the "outsourcing of refugee processing" (Stephens 2019). Would-be refugees were required to apply and be approved before they enter the EU, while various transit countries are subsidized by the EU to house them while their applications are pending (Oudejens et al. 2018; Stevis-Gridneff 2019). A number of states to the south of the EU became what might be called "holding states" where growing numbers of would-be refugees or other migrants were held in camps while their applications were considered. A second feature is the prohibition of "forum shopping" for asylum claims. Asylum seekers, if not processed before arrival, are required to apply for and be processed in the EU state that they first enter, rather than traveling within the EU to find a more favorable forum (Dublin III Regulation 2013). Approval rates for asylum applications vary widely from nation to nation within the EU (Eurostat 2018).

5 The Breakdown of the EU Asylum Regime in 2015

Migration, licit and illicit, into and within the European Union, had caused growing resentment towards newcomers well before 2015 (Hall 2000). Intensifying violence in Afghanistan and Syria, the rise of ISIS in Iraq and Syria, and droughts and other problems across West and Central Africa, at first did not produce a major crisis for Europe because "holding states" formed a barrier against political and economic refugees reaching Europe. The "Arab Spring" caused the breakdown of Libya and Tunisia as holding states, allowing a flood of migrants (refugees or otherwise) to cross the Mediterranean, primarily to Italy and Malta (UNCHR 2015). Then Recep Erdoğan decided to allow refugees held in Turkey to enter Europe as a means of pressuring the EU on issues including, among others, funding for holding the refugees in Turkey (Winsor 2016). Europe was flooded with refugees from the Middle East and Africa. The total number of persons applying for refuge in Europe jumped from 300,000 in 2014, to 1,300,000 in 2015 (Pew Research Center 2016). Another 1,200,000 sought refuge in 2016 (BBC News 2018).

The EU migration policy collapsed under those numbers. Most eastern European countries closed their borders to new refugees, ignoring their own not so distant need for refugee protections for their citizens (Lyman 2015). In doing so, they violated their obligations under the 1951 *Convention* and under EU law, and when the EU Council of Ministers sought to relieve the burden on any one state by creating quotas for the number of refugees each was to accept, the eastern European states refused to accept their quotas (Kanter 2017). All of this was notwithstanding that once an applicant's refugee status was approved, he or she could move elsewhere.

In contrast, Angela Merkel, Chancellor of Germany, recalling German's prior culpability in creating refugees during and after World War II, announced that Germany would open its borders to all refugees (Holehouse et al. 2015). The other European countries directed refugees to Germany so that approximately 1,200,000 asylum applications were received in Germany in 2015 and 2016, out of about 2,500,000 for the entire EU for those years (Bennhold et al. 2015). This produced a political backlash against Merkel, leading to the rise of the Alternative for Germany (AfD) Party (Eddy and Bennhold 2019). In didn't help that some small percentage of refugees committed highly visible crimes, discrediting all refugees in the eyes of many (Eddy 2019).

In Italy, the number of asylum applications surged in 2016 to 123,000, putting Italy in second place, well behind Germany (AIDA 2017). The number of asylum seekers in all Europe fell back to 300,000 after the EU agreed to pay Turkey's expenses for holding migrants in Turkey (Winsor 2016). Yet even the figure of 300,000, widely considered tolerable before 2015, now seemed intolerable (Kingsley 2018). Instead, demagogues across the continent used fear of refugees (and other migrants) to consolidate their power, with Brexit serving as only the most extravagant example (Brubaker 2017). Erdoğan notably threatened to release the migrants again as a response to European criticism of his attack on the Kurds in northern Syria in 2019 (Gall 2019).

The deal with Erdoğan did not affect the flow of people across the Mediterranean from Libya and Tunisia, so the presence and treatment of migrants (whether refugees or otherwise) became a prominent feature of Italian politics after 2015 (Horowitz 2019b). Under the leadership of Matteo Salvini, Minister of the Interior and leader of *la Liga,* Italy imposed severely tightened restrictions on who could apply for or receive asylum there—in clear violation of Italian laws and EU regulations. As of 2019, only three narrow classes of people were considered for asylum in Italy: persons who had suffered labor exploitation in Italy; victims of human trafficking; and people fleeing natural disasters (Senato della Repubblica 2018). Instead, Italy (and other Mediterranean states) undertook to intercept boats of migrants at sea to bar their landing in the EU and force them back to North Africa (Specia 2019), often with tragic results (Momigliano 2019; Momigiano and Minder 2019). Whether this situation will change with the (at least temporary) fall of Salvini from power remains to be seen (Horowitz 2019a).

6 The Duty to Save Lives at Sea

Italy's government refuses to allow derelict vessels to land if they carry migrants, whether refugees or not. As a result, significant numbers of people drown in the Mediterranean trying to reach Europe (Momigliano 2019). Italy is ignoring its duties under the 1951 *Convention* and under EU and Italian law, but arguably doesn't actually violate those duties if the people are not yet present Italy. This posture does, however, definitely violate Italy's duties regarding saving lives at sea.

The recognition of a formal duty to save lives at sea goes back to the sinking of the *HMS Titanic* in 1912. Before then, the saving of lives on another ship was entirely discretionary with the captain of any vessel near the endangered vessel. The major maritime states came together in 1913 to draft the *Convention for the Safety of Life at Sea* (SOLAS 1974). That treaty and subsequent legal developments, now found in the *UN Convention on the Law of the Sea,* require nearby ships and coastal states to save lives endangered at sea "with all possible speed" (UNCLOS 1982, art. 98(1)). Ships are excused from their duty to rescue only for lack of ability or necessity or if it creates an unreasonable risk. Furthermore, coastal states are required to maintain and operate search and rescue operations (art. 98(2)). The duty then falls on the rescuer to deliver the rescued persons to a place of safety—dry land.

There is no duty on the part of a coastal state to receive the rescued persons—they could be delivered to another willing coastal state, so long as the delivery does not amount to *refoulement* (1951 Convention, arts. 32, 33). The *International Convention on Maritime Search and Rescue* does require coastal states to allow rescuing ships to be relieved of their burdens with "a minimum of further deviation" from their voyages (SAR Convention 1979, ch. 3.1.9). There is no ambiguity on whether these duties apply to Italy when it is the nearest coastal state. Italy has incorporated the life-saving duties into its Navigation Code (Code 1942, art. 498). The Italian arrest of the captain of a ship rescuing migrants in danger at sea leaves no doubt about Italy's violation (Provoledo 2019a). While a judge ordered the release of the captain after about a month in jail (Provoledo 2019b), that hardly diminishes the violation.

7 Conclusions

Today's refugee flows are not a new phenomenon, not even in terms of numbers. Elaborate legal structures, including the 1951 *Convention,* EU laws and regulations, and national laws, were created to deal with such problems. Nonetheless, many states are systematically violating those structures (Horowitz 2019c; Pijnenburg 2018). In the face of populist demands for "sealing the borders," those laws are likely to remain unenforced.

For violations of a treaty, the usual remedy is "international responsibility"—the violating state is to make amends to other states party to the treaty. Which state is injured by a violation of the 1951 *Convention?* Not the state from which the would-be refugees are fleeing. That state doesn't care what happens to people whom it made flee. Other parties to the convention? If they care so much, they just take in the refugees. So which state then will complain about the violation? Similarly, it is abundantly clear that the EU lacks either the will or the means to enforce its refugee policies. National laws, like Italy's, can be changed by a "populist" regime—and no doubt will be if a court sought to enforce the existing law, whether national legislation or international obligation. If such violations of refugee policy are "widespread and systematic attacks" on a civilian population, those responsible for the violations could be prosecuted under the *Statute of the International Criminal Court* for crimes against humanity (Rome Statute 1998 art. 7), assuming the jurisdictional prerequisites for the Statute apply.

All in all, the prospects for protecting the rights of refugees in Europe do not appear good.

References

1951 Convention UN Convention on the Status of Refugees and 1967 Protocol. Available at https://www.unhcr.org/en-us/3b66c2aa10
Advisory Opinion (2007) Advisory Opinion on the Extraterritorial Application of the *Non-Refoulment* Obligation under the 1951 Convention. International Court of Justice, 26 January
Ahmed A, Villegas P (2019) 'This takes away all hope': rule bars most applicants for asylum in U.S. NY Times, 13 Sept. A18
AIDA (2017) Asylum Information Database, Country Report: Italy, 2016 Update. Available at file:///C:/Users/jwdel/Downloads/aida_it_2016update.pdf
Antons J-H (2014) Displaced persons in post-war Germany: parallel societies in a hostile environment. J Contemp Hist 49(1):92–114
Bar M, Leukhina O (2010) Demographic transition and industrial revolution: a macroeconomic investigation. Rev Econ Dyn 13:434–451
BBC News (2018) Migration to Europe in Charts. 11 Sept. Available at https://www.bbc.com/news/world-europe-44660699
Bennhold K et al (2015) Germans welcome migrants after long journey through Hungary and Austria. NY Times, 6 Sep., p A1
Bodor DP (1950) Topical autobiographies of displaced people recorded verbatim in displaced persons camps with a psychological and anthropological analysis. s.n., Chicago
Brubaker R (2017) Why populism? Theory Soc 46:357–385
Bull S (2014) Trench: a history of trench warfare on the western front. Osprey Publishing, Oxford
Carpenter K, Mansel P (eds) (1999) The French Emigrés in Europe and the struggle against revolution, 1789–1814. St. Martin's Press, New York
Cave D (2019) A timeline of despair in Australia's Offshore Detention Centers. NY Times, 28 June, p A5
Chin R (2007) The guest worker question in postwar Germany. Cambridge University Press, Cambridge
Clark B (2009) Twice a stranger. Harvard University Press, Cambridge
Clark L (2016) Blitzkrieg. Atlantic Monthly Press, New York
Clark JE Jr (2004) Railroads in the Civil War. LSU Press, Baton Rouge
Code (1942) Codice della Nagicione (Italy)
Conquest R (1968) The great terror. MacMillan Co, New York
Conquest R (1986) The harvest of sorrow. Oxford University Press, Oxford
Dikötter F (2016) The cultural revolution. Bloomsbury Press, New York
Douglas RM (2013) Orderly and humane: the expulsion of the Germans after the second world war. Yale University Press, New Haven
Dublin III Regulation (2013) Regulation (EU) No. 604/2013, 26 June, OJ L180/31
Eddy M (2019) A summer of angst in one of the world's safest countries. NY Times, 14 Aug., p A10
Eddy M, Bennhold K (2019) German far right makes election gains, but falls short of victory. New York Times, 2 Sep., p A10
Ehrlich MA (ed) (2018) Encyclopedia of the Jewish diaspora. ABC-CLIO, Santa Barbara

Eurostat (2018) Asylum Quarterly Report. Available at https://ec.europa.eu/eurostat/statistics-explained/index.php/Asylum_quarterly_report#Decisions_on_asylum_applications
Fox G (1950) The origins of UNRRA. Polit Sci Q 65:561–584
Fraser A (1995) The gypsies, 2nd edn. Wiley-Blackwell Publishers, Hoboken
Gall C (2019) Turkey's radical plan: send a million refugees back to Syria. NY Times, 11 Sep., p A4
Genocide Convention (1948) Convention on the Prevention and Punishment of the Crime of Genocide, 9 Dec. 1948, entered into force 12 Jan. 1951. Available at https://www.un.org/en/genocideprevention/documents/atrocity-crimes/Doc.1_Convention%20on%20the%20Prevention%20and%20Punishment%20of%20the%20Crime%20of%20Genocide.pdf
Hall B (2000) Immigration in the European Union: problem or solution? OECD Observer. Available at http://oecdobserver.org/news/archivestory.php/aid/337/Immigration_in_the_European_Union:_problem_or_solution_.html
Harvati K et al (2019) Apidima cave fossils provide earliest evidence of *Homo Sapiens* in Eurasia. Nature 571:500–504
Hassell JE (1991) Russian refugees in France and the United States between the world wars. Trans Am Philos Soc 81(7):1–96
Hatton TJ, Williamson JG (1994) What drove the mass migrations from Europe in the late nineteenth century. Popul Dev Rev 20:533–559
Hermann P (1954) Conquest by man. Harper & Bros, New York
Hinton AL et al (2014) Colonial genocide in indigenous North America. Duke University Press, Durham
Holehouse M et al (2015) Germany drops EU rules to let in Syrian refugees. The Telegraph, 24 Aug. Available at https://www.telegraph.co.uk/news/worldnews/europe/germany/11821822/Germany-drops-EU-rules-to-allow-in-Syrian-refugees.html
Horowitz J (2019a) Europe tries to seize momentum for new deal on migration. NY Times, 23 Sept. A11
Horowitz J (2019b) In Matteo Salvini's Italy, good is bad and 'do-gooders' are the worst. NY Times, 15 April, p A10
Horowitz J (2019c) Salvini's standoff at sea highlights Italy's war on rescue ships. NY Times, 17 Aug., p A9
IRO Constitution (1946) Constitution of the International Relief Organization, 15 Dec., 18 UNTS 387
Kanter J (2017) E.U. states must accept their share of migrants, court rules. NY Times, 7 Sep., p A10
Kévorkian R (2011) The Armenian Genocide. I.B. Tauris & Co, New York
Kiernan B (2008) The Pol Pot Regime, 3rd edn. Yale University Press, New Haven
Kingsley P (2018) Migration to Europe is down sharply. So is it still a crisis? NY Times, 27 June. Available at https://www.nytimes.com/interactive/2018/06/27/world/europe/europe-migrant-crisis-change.html
Kingsley P (2019) Is Trump's America tougher on asylum than other western countries? NY Times, 15 Sept., p A18
Kogon E (1950) The theory and practice of hell. MacMillan Co, New York
Laqueur W (ed) (2001) The holocaust encyclopedia. Yale University Press, New Haven
Lyman R (2015) Eastern bloc's resistance to refugees highlight's Europe's cultural and political divisions. NY Times, 13 Sep., p A1
MacKensie SF (1994) The treatment of prisoners of war in World War II. J Mod Hist 66:487–520
McPherson SS (2019) The global refugee crisis: fleeing conflict and violence. Twenty-First Century Books, Minneapolis
Momigiano A, Minder R (2019) Migrant ship stranded off Italy in 'crisis,' Aid Group Says. NY Times, 19 Aug., p A7
Momigliano A (2019) About 150 migrants drown in shipwreck off the coast of Libya. NY Times, 26 July, p A9
Murphy HBM (1952) Assimilating the displaced person. Aust Q 24:46–59
Oudejens N et al (2018) Protecting the EU external borders and the prohibition of refoulement. Melb J Int Law 19:1–25

Pace A (1974) Mexican refugees in Arizona, 1910–1911. J Southwest 16:5–18

Pew Research Center (2016) Number of refugees to Europe surges to record 1.3 million in 2015. 2 Aug. Available at https://www.pewresearch.org/global/2016/08/02/number-of-refugees-to-europe-surges-to-record-1-3-million-in-2015/

Pijnenburg A (2018) From Italian pushbacks to Libyan pullbacks: is *Hris 2.0* in the making in Strasbourg? Eur J Migr Law 20:396–426

Pohl OJ (1999) Stalin's ethnic cleansing. Greenwood Press, Westport

Provoledo E (2019a) Italy arrests captain of ship that rescued dozens of migrants at sea. NY Times, 30 June, p A4

Provoledo E (2019b) Judge in Italy orders release of captain of migrant-rescue ship. NY Times, 3 July, p A4

Rome Statute (1998) Rome Statute of the International Criminal Court, 17 July. Available at https://www.icc-cpi.int/resource-library/Documents/RS-Eng.pdf

SAR Convention (1979) International Convention on Maritime Search and Rescue, 17 April, UNTS 1405:119–256

Schabas W (2000) Genocide in international law: the crimes of crimes. Cambridge University Press, Cambridge

Schechtman J (1947) European population transfers, 1939–1945. Harvard University Press, Cambridge

Senato della Repubblica (2018) Attesto che il Senato della Repubblica, il 7 novembre 2018, ha approvato il seguente disegno di legge, d'iniziativa del Governo. 7 Nov. Available at http://www.senato.it/service/PDF/PDFServer/BGT/01080098.pdf

SOLAS (1974) The International Convention for the Safety of Life at Sea, originally 20 Jan. 1914, as updated on 1 Nov. 1974. Available at http://www.imo.org/en/About/Conventions/ListOfConventions/Pages/International-Convention-for-the-Safety-of-Life-at-Sea-(SOLAS),-1974.aspx

Specia M (2019) Hundreds of migrants stranded in Mediterranean in stand off over aid ships. NY Times, 13 Aug., p A7

Statute of the UNCHR (1950) Statute of the UN High Commissioner for Refugees, 14 Dec., A/RES./428(V)

Stephens B (2019) An immigration policy worse than Trump's. NY Times, 6 July, p A17

Stevis-Gridneff M (2019) Europe keeps asylum seekers at a distance: this time in Rwanda. NY Times, 8 Sept., p A8

Tóth H (2014) An exiled generation: German and Hungarian refugees of revolution, 1848–1871. Cambridge University Press, Cambridge

Treaty of Amsterdam (1997) Treaty of Amsterdam Amending the Treaty on the European Union, 2 Oct., OJ 97/C 340/01

UNCHR (2015) The sea route to Europe: the Mediterranean passage in the age of refugees. Available at https://www.unhcr.org/5592bd059.html#14

UNCHR (2019) UNCHR figures at a glance. Available at https://www.unhcr.org/en-us/figures-at-a-glance.html

UNCLOS (1982) United Nations Convention on the Law of the Sea, 10 Dec. Available at https://www.un.org/depts/los/convention_agreements/texts/unclos/unclos_e.pdf

UNRRA Agreement (1943) Agreement for United Nations Relief and Rehabilitation Administration, 9 Nov. Available at http://www.ibiblio.org/pha/policy/1943/431109a.html

von Plato A et al (eds) (2010) Hitler's slaves. Berghahn Press, New York

Walsh D (2019) Airstrike kills dozens of migrants at detention center in Libya. NY Times, 4 July, p A9

Winsor M (2016) EU-Turkey Summit: Ankara uses refugee crisis as leverage for unexpected demands. International Business Times, 7 May. Available at https://www.ibtimes.com/eu-turkey-summit-ankara-uses-refugee-crisis-leverage-unexpected-demands-2332065

Wyman M (1998) DPs: Europe's displaced persons, 1945–1951. Cornell University Press, Ithaca

Part II
Public Health and Ecointegrity

Using Science and Law to Promote Humans' Biological Integrity

Carl F. Cranor

Abstract This chapter first briefly considers the idea of humans' biological integrity and a characterization of chronic diseases, the dominant source of illness in developed countries. Next, it reviews some highlights of chronic disease risks to the most vulnerable of humans, children, and, finally, considers how both science and the law can fail to protect the public, or in contrast how they can protect the public much better than at present.

1 Human Biological Integrity

Biological integrity is typically seen as "the inviolability of the physical body and emphasizes the importance of personal autonomy and the self-determination of human beings over their own bodies" (Findwords). I disregard much of this highly normative account, focusing only the "inviolability of the physical body." I gloss this with the idea that people have biological integrity provided they have sufficiently good health to flourish in their society as the thinking, goal-oriented creatures humans are (compared to less endowed biological entities such as sheep or birds) (Daniels 1981).

Serious chronic diseases or dysfunctions, can substantially disrupt or burden humans' opportunities (relative to the community of which they are a part) to pursue careers, life plans, and a full range of prospects into old age for which they have appropriate talent, abilities and motivations (Fries 1980; Daniels 1981; Cranor 2017a).

C. F. Cranor (✉)
University of California, Department of Philosophy, Riverside, CA, USA
e-mail: carl.cranor@ucr.edu

L. Westra et al. (eds.), *Ecological Integrity in Science and Law*,
https://doi.org/10.1007/978-3-030-46259-8_6

2 Chronic Diseases

Chronic diseases are contrasted with the more common infectious diseases that can be transmitted directly from other persons, insect or animal bites, or indirectly from contamination in the environment, typically via bacteria and viruses (Mayo Clinic). For centuries infectious diseases have ravaged humans with common diseases along with flu and plague epidemics.

About 100 years ago public health officials began reducing and often effectively eliminating infectious diseases in the United States and other developed countries. They cleaned up polluted environments, such as horse manure in streets and sewage dumped into rivers in which children swam and from which communities took their drinking water, and chlorinated drinking water. Finally, researchers developed antibiotics to treat bacterial diseases and vaccines to prevent viral disorders, as they came to understand these sources of harm. These combined efforts left chronic diseases as the major source of maladies experienced by people in developed countries (Armstrong et al. 1999).

Chronic diseases often originate in early life and develop insidiously. They persist "for a long time … [typically] lasting 3 months or more …. [and] generally cannot be prevented by vaccines or cured by medication, nor do they just disappear" (MedicineNet). They are now the leading causes of morbidity and mortality in developed countries (Fries et al. 2011; Vineis et al. 2014). They can result from bad luck, unfortunate genes, personal behavior, or *the actions of others*, and can interfere with normal biological functioning, good health, and human flourishing.

Physicians are concerned, *inter alia*, with arteriosclerotic plaque that can lead to acute thrombotic events or vascular insufficiency, osteoarthritic bone spurs that trigger joint pain or disability, glucose intolerance resulting in diabetes, and emphysema that can reduce pulmonary reserves. Neoplasms become cancerous with age (Fries et al. 2011).

Physicians recommend patients avoid behaviors that risk causing the diseases: never smoke, avoid obesity, avoid sedentary behavior, and don't drink to excess (Fries et al. 2011). Adhering to this advice can delay "the age of onset of chronic illness [with its morbidities]… and [squeeze] the most of [life's] morbidity into a shorter period with less lifetime disability" (Fries et al. 2011).

Failure to make health protective choices may lead to lesser or more serious morbidities that may interfere with normal biological functioning and to greater or lesser misery during one's lifetime. Individual choices plus medical care can reduce the risks of many common chronic diseases. Fries et al. (2011) explain that an important facilitator of chronic diseases is having insufficient organ reserves to help withstand the onset of diseases. People who exercise consistently can have eight to sixteen years of lesser morbidities compared with average non-exercisers.

Importantly for what follows chronic diseases can also be caused by the actions of others in creating and distributing substances in commerce that are toxic to human beings. Toxic substances can cause *inter alia* neurological dysfunctions, immune system disorders, cancers, lung diseases, some coronary-artery diseases and some diabetes (Trasande and Liu 2011).

However, diseases caused by the products or actions of others (versus those resulting from personal choices) pose issues of justice. (1) Chronic diseases and their morbidities actually cause harm to those who suffer from them. (2) They can also be unwelcome surprises, disrupting and arbitrarily interfering with opportunities (compared with those available to typical healthy members of the community) over a complete lifespan (Daniels 1981; Cranor 2017a). A powerful principle of justice—the "fair equality of opportunity principle"—identifies what is wrong with interfering with opportunities.

A rationale for this principle is that every country (developed or less developed) has a range of opportunities for its citizens depending upon its wealth, technological and industrial advances, education, natural resources, and so on. It is well settled that it is unjust to legally exclude people from opportunities in a society, e.g., by enacting laws forbidding women, disfavored minorities or those with particular religions from pursuing life goals compatible with their talents abilities and motivations (Rawls 1999). There are more subtle barriers to opportunity and reasons of justice support the idea that no citizen should be arbitrarily disadvantaged from pursuing his or her opportunities consistent with that person's talents, abilities and motivation, even though some will be more fortunate and some less fortunate in the distribution of life's talents, abilities and motivations (Daniels 1981; Rawls 1999). This recognizes the equal status and standing of each person in the community.

Educational and medical institutions have a role in both minimizing more subtle arbitrary disadvantages that can befall citizens as well as enhancing opportunities for those who may experience subtle barriers to opportunity. These institutions should provide for *everyone* in the community, regardless of talents, abilities, and motivations, to have a fair opportunity to develop his or her endowments to the best of his/her ability in order to achieve life goals conferred by his or her particular bundle of endowments.

Both institutions are strategically important for counteracting obvious and less visible barriers that arbitrarily and subtly can reduce one's chances of a flourishing life within the range of a country's opportunities.

Because some chronic diseases result from personal choices Fries and others recommend that physicians urge their patients to make personal choices that *primordially prevent* risks that lead to chronic diseases, which can bring associated morbidities that substantially interfere with their life long opportunities (Fries et al. 2011). Similarly, legal institutions can be chosen and designed to primordially reduce risk factors that lead to chronic diseases caused by the products or actions of others.

Thus, the fair equality of opportunity principle requires that both medical care and legal institutions should seek to (a) *prevent diseases* with good choices, timely medical treatment, and environmental health protections; (b) *treat diseases* that undermine persons' health by ensuring access to medical care, and (c) m*aintain* the chronically ill as close to normal functioning as their ages and conditions allow by providing medical treatment and accommodations (Daniels 1981; Cranor 2017a). The remainder of this discussion focuses only on (a), just legal protections from toxic substances.

3 The Law's Contribution to Health Protections

Administrative health laws can prevent/reduce toxicity-caused chronic diseases, while personal injury (or the tort law) can support treatment and redress for diseases caused by others (but I do not discuss this).

Administrative health laws exhibit two generic strategies that might protect people in a community from diseases caused by toxic substances.

In the U.S. laws for drugs (1962) and pesticides (1968) require products *to undergo routine* premarket toxicity testing with a battery of tests and *agency review* before they can enter commerce (U.S. Congress 1987). These laws cover only about 10–20% of chemical creations. If premarket laws function well, they approximate the primordial prevention of risks of diseases.

Postmarket laws seek to protect citizens from toxicants among general chemicals by permitting chemical creations to enter commerce with *no legally* required routine testing or approval under the 1976 Toxic Substances Control Act. This covers about 80–90% of chemical creations. Any protection from this law depends in part upon whether the U.S. Environmental protection Agency (EPA) is able to discern toxic properties of products entering commerce when it has minimal to nearly non-existent information provided by a chemical manufacturer (toxicity data is not required) and then asking for additional data if it seems needed to ensure a product's safety. The law also assumed that risk assessments post-exposure could be quickly conducted to identify toxic risks so they could be reduced or removed before (too many) people were harmed. These legal assumptions were clearly mistaken (Cranor 2017a). Postmarket laws have permitted citizens to be contaminated by more than 300 toxic substances (U.S. Department of Health and Human Services, CDC 2019). Pregnant women are contaminated with up to forty-three toxic substances (Woodruff et al. 2011) and children are born with toxicants in their bodies (Cranor 2011).

Moreover, identifiable people have been harmed and had their lifetime opportunities truncated or eliminated altogether because of exposures to toxicants. At age 47 Brian Milward contracted a rare form of leukemia, promyelocytic leukemia, caused by benzene, a known carcinogen, used as a solvent in fluids he applied as a refrigerator repairman (Milward v. Acuity Specialty Products 2011). His disease, chemotherapy, diabetes, and a rare bowel disorder left him at age 57, "with 'absolutely ridiculous' fatigue" (Lombardi 2014). He had to retire and take disability with greatly diminished opportunities. He couldn't do what he loves: repair racecars, work in his yard, or even play with his grandchildren. "It just sucks when you get a cancer like this," he said (Lombardi 2014). The companies creating products with benzene and the U.S. Occupational Safety and Health Administration failed to protect him from harm, substantially reducing his lifetime opportunities.

Hairdresser Sandy Guest used Brazilian Blowout to straighten her clients' hair. It was "loaded with formaldehyde" and she died of leukemia at age 55 (Morris 2015b). Her death was a preventable tragedy: from 1981 to 2011 seventeen human studies revealed that formaldehyde caused nasopharyngeal and sinonasal cancers, with some showing it caused myeloid cancers, but no federal agency protected citizens from formaldehyde. Industry resistance to regulation and slow agency review likely led to Guest's death (Cranor 2017a).

Similarly, individuals who merely drank tap water contaminated with the "forever chemical" PFOA (Perfluorooctanoic acid) or C8 used in Teflon, Gore-Tex, and flame retardants, *inter alia,* contracted kidney and testicular cancer. Although they received tort law compensation for their injuries, their serious diseases, associated harms, and monetary costs truncated lifetime opportunities (Mancini 2017). DuPont employees suffered similar diseases (Lerner 2015; Cranor 2017a). DuPont and administrative institutions failed all these people, although the tort law provided some compensation for some of their losses (Mancini 2017).

4 The Developmental Origins of Disease

Postmarket laws did not protect adults from chronic diseases, but a new science—the "developmental origins of disease"—greatly heightens the urgency to better serve this goal (Grandjean et al. 2008).

Children are among the most vulnerable humans exposed to toxicants (recall exposures above). "*In utero* nutrition and/or *in utero* or neonatal exposures to toxicants alter susceptibility to disease later in life and … [can] result in death, malformations, low birth weight or functional changes including increased susceptibility to diseases later in life." (Heindel 2008). This shows chronic diseases can originate from environmental insults to embryos to fetuses to infants to teenagers (Grandjean et al. 2008; Cao et al. 2016), even to later life stages (Heindel 2018). For instance, women are particularly vulnerable to breast cancer from toxicants during puberty (Cohn et al. 2007). The diagram below (see Fig. 1), based on animal data, suggests people are not merely exposed once, but also likely face continuous exposures at different susceptibility stages from possibly interacting substances over a lifetime (NRC 2009; Heindel 2018).

During development children's organ systems are more susceptible to toxicants. *In utero,* during breast-feeding, and as infants they have greater exposures to any toxicants per body weight than adults. Their biological defenses are less well developed to protect them from invaders. And, they have a longer lifespan for diseases to develop, with adverse effects to some organ systems being irreversible, e.g., the brain, the immune system, and likely the reproductive system (Grandjean et al. 2008; Cranor 2017a).

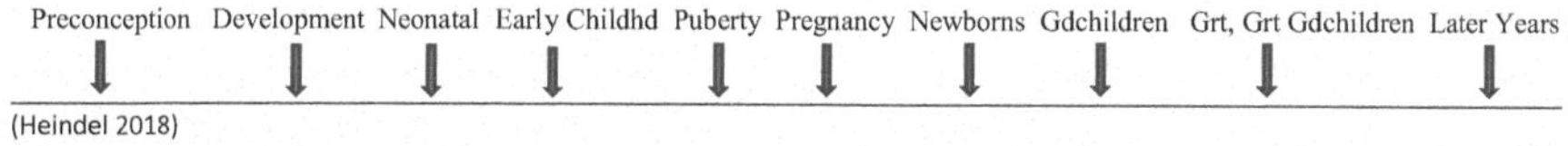

Fig. 1 Timeline of vulnerable life stages in a person and her offspring

Three disease catastrophes signaled wider and more consequential problems with toxic exposures: *in utero* exposures to methylmercury (1950s), thalidomide (1950s and 1960s) and diethylstilbestrol (1950s–1960s) (Cranor 2011, 2017a). There is "no placental barrier *per se:* the vast majority of chemicals given a pregnant animal (or woman) reach the fetus in significant concentrations soon after administration." (Schardein 2000). Recently created plastic nanoparticles can also cross the placenta, contaminating and potentially harming embryos or fetuses (Wick et al. 2010). Tiny doses of mutagenic carcinogens, lead, Thalidomide, and some other compounds can trigger diseases (Cranor 2017a). And all of this may adversely affect certain subsets of individuals because of identified susceptibility genes for polycyclic aromatic hydrocarbons (Perera et al. 1999), organophosphate pesticides (Huen et al. 2009) and methylmercury (Julvez et al. 2013).

Moreover, these are not merely theoretical points. Trasande and Liu (2011) estimate that the *annual* costs of pediatric diseases of environmental origin include the following: neurological conditions caused by lead ($50.9 billion) and methylmercury ($5.1 billion), intellectual disability ($5.4 billion), autism ($7.9 billion), attention deficit hyperactivity disorder ($5.0 billion), along with the immune system disorder of asthma ($2.2 billion), and childhood cancer ($95 million). A best total estimate of these diseases is $76.6 billion (see also Cranor 2017a).

Animal data reveal even greater concerns for ill-timed exposures: in utero exposures to toxicants can trigger transgenerational reproductive and other disorders. *In utero* toxic exposure when reproductive organs are developing causes transgenerational reproductive harm and some cancers in males and females alike through four generations (Cranor 2017a).

5 Occupational Risks to Adults and Children

Finally both parents contaminated with toxicants, not just women, can adversely affect developing children. Males contaminated with Paxil, anesthetic gases, morphine, lead, mercury, pesticides, solvents, dyes, and paints can produce miscarriages, along with prenatal or neonatal problems (Cranor 2017a).

Contaminated parents raise urgent issues for occupational protections. Workers are typically "callously" unprotected (Morris 2015a). For instance, "Yvette Flores's body was a 'toxic warehouse before [her son] Mark was conceived.'" He was born with "extensive cognitive impairment," caused by *in utero* lead exposure (Morris 2015a) that violated his bodily integrity. At age 36 he has no ordinary opportunities available to the vast majority of fellow citizens; he will need the care of an adult the rest of his life because he cannot conduct daily living activities on his own. This points to yet another issue of justice: occupational settings should have increased (primordial?) occupational protections from risks to adults *and their future children* (Cranor 2017a).

Postmarket laws are unjust; they cannot prevent chronic *diseases and morbidity* in children or adults caused by others, permitting harm to both and undermining lifetime opportunities for some of us. The developmental origins of disease and our permeability to toxicants increase the urgency for better public health protections. There are no analogues to Harry Potter cloaks against toxicants or for preventing developmental vulnerability. Vaccines and antibiotics, two public health tools that helped eliminate infectious diseases, won't prevent or treat them.

6 Institutional Interventions

Two institutional interventions are critical for protecting the public: one broadly legal, the other scientific. Administrative health laws must primordially prevent risks from toxic exposures, "cleaning up" toxicants as public health officials once cleaned up streets, sewage, water, and the air. Premarket testing and approval laws with routine toxicity testing and agency approval are needed for all created chemical products that expose the public (Cranor 2011, 2017a).

First, the toxicity of chemical products must be understood before they enter commerce, exposing the public. The 1976 Toxic substances Control Act failed on this dimension because substances were permitted into commerce with little or no toxicity data. Second, once products in commerce are revealed as toxic, agencies must more quickly identify and reduce risks. Administrative agencies have been quite slow to remove toxic products from commerce. Premarket laws address the first issue, while both premarket and postmarket laws must more quickly reduce the risks of toxic products once they have been identified.

Second, science has an important role in assisting the identification of toxicants. How much science should be required to support a claim that a proposed new substance will likely not pose risks to the public? Once a product is in commerce, how much science under a law should be needed to support the reduction of risks or the removal of the product from commerce. I largely focus on this latter issue because toxic products have been so difficult to remove from commerce. For example, under the 1976 TSCA, the U.S. EPA was not permitted to remove all uses of asbestos from commerce despite 45,000 pages of scientific and legal support and a decade of effort; a court precluded this action (Corrosion Proof fittings v. Environmental Protection Agency 1991). If the EPA cannot remove one of the most toxic substances from commerce, there is little hope for removing others. In addition, the EPA has had in its queue for regulation four known or probable carcinogens (along with several other substances) for one or more decades with numerous fits and starts along the way, e.g., trichloroethylene, dioxin, perchloroethylene, and formaldehyde (GAO 2008).

7 Some Scientific Policies and Research That Assist Quicker Assessments of Toxicants

Some scientific procedures can more rapidly identify toxicants and, thus, better protect the public. Administrative agencies could make greater use of well-founded scientific *defaults* or *presumptions* to assist in their tasks. Typically such defaults are based on empirical evidence and/or important scientific or social values identifying a favored position—e.g., protection of children, using linear extrapolations to identify the carcinogenic potential of a substance—or a disfavored view—e.g., toxicants having related features should be treated similarly (Cranor 2018).

Defaults or presumptions create provisions for departing from the status quo ante, identifying existing or new assumptions to create defaults, and avoiding "full *ad hoc* examination of data and the [full] spectrum of inferences…" (NRC 2009; Cranor 2018). They provide starting places, clarify discussions, avoid ad hoc choices, facilitate predictable and consistent risk assessments, and provide a basis for the public assessment actions (NRC 2009; Cranor 2018).

For instance, the Food Quality Protection Act mandates at least a ten-fold safety factor for toxicants in food (NRC 2009). Second, positive findings in animal cancer studies indicate that an agent can have human carcinogenic potential (NRC 2009). Third, "[T]he basic pathways of metabolism and the occurrence of metabolites in tissues in … species-to-species extrapolation of cancer hazard and risk" are similar (NRC 2009). Fourth, linear extrapolations from high doses to low dose effects are health protective, "When the weight of … all available data are insufficient to establish the mode of action for a tumor site (NRC 2009). There are numerous others, but the above examples suggest the idea.

Presumptions can facilitate assessing and removing toxicants from commerce. For example, the International Agency for Research on Cancer (IARC) and others found that vinyl chloride (VC) is highly toxic to humans based on occupational exposures in polyvinyl chloride plants (Heath et al. 1975). The chemically similar compounds vinyl fluoride (VF) and vinyl bromide (VB) act by similar biological mechanisms. Thus, IARC listed VF and VB as "probable human carcinogens" without having statistically significant evidence that VF and VB causes cancer in humans; knowledge of the toxicity of VC, the results of animal studies, and mechanistic similarities of VF and VB to VC were sufficient for the inferences (Cogliano et al. 2008).

In addition, IARC used mechanistic data to upgrade six probable human carcinogens to known human carcinogens and to upgrade thirty-nine substances to probable human carcinogens (Cogliano et al. 2008).

Presumptions can facilitate quicker identification and ultimately, removal of products from commerce because of increased scientific understanding of biochemical interactions. Of course, once there is such a body of knowledge it can also be used in premarket contexts to better identify products as toxic before they enter commerce and put the public at risk.

A final scientific advance meriting comment is a series of recent papers identifying "key characteristics" of toxicants leading to disease. The idea is to identify "upstream" indicators that portend the development of disease and combine those with other knowledge so that one does not need to wait for a full fledged disease to show up in a population (Smith et al. 2016).

Both facilitating quicker removal of products from commerce and having the same scientific tools to identify toxic substances before they even enter commerce better protect the public. These favorable scientific developments hold the promise of better protection of the public and workforce in the future, but I do not want to be overly optimistic in the face of substantial industry pressures.

8 Going Forward to Better Protect the Public

The U.S. and likely most countries have a huge backlog of poorly understood chemical products in commerce. *Fifty years ago* the U.S. President's Council on Environmental Quality called attention to this problem and the need for better laws:

> We should no longer be limited to repairing the damage after it has been done [Torts]; nor should we continue to allow the entire population or the entire environment to be used as a laboratory [postmarket] (US Council on Environmental Quality 1971)

Because the 1976 TSCA permitted the creation of 84,000 substances whose toxicity is poorly understood, this is about where we find ourselves in 2019.

The U.S. Congress in 2016 finally acknowledged this issue, passing the Frank R. Lautenberg Chemical Safety for the 21st Century Act, which seeks to address some of these concerns. For new chemicals the EPA "must make an affirmative finding on the safety of a new chemical or significant new use of an existing chemical before it is allowed into the marketplace" (EPA 2016). For both new and existing chemicals, it seeks to protect susceptible people: children, highly exposed workers, pregnant women, and the elderly. Finally, it has "rigorous" postmarket review deadlines so that products do not become bogged down because of company recalcitrance and endless scientific and legal disputes (Cranor 2017a). These are not the only provisions, but they might represent important improvements for protecting the public's health if they are well administered in the spirit of the law.

However, there is the rub. Will the evidentiary bar for *new* substances be set too low, permitting products into commerce with too little data and poor understanding of their toxicity? This would recapitulate the 1976 TSCA. The Trump administrations' approval in 2017 of 600 new substances in a few months (an unheard of rate) creates the impression that public health protections were not a high value for this administration and is quite different from the review of pharmaceuticals and pesticides.

Will the evidentiary bar for *existing* substances be too high? EPA's refusal to ban chlorpyrifos, a pesticide with an overwhelming body of data about its toxicity, and against the protests of EPA scientists also does not bode well for protecting the general public and susceptible subpopulations (Cranor 2017b).

These examples and the previous discussion show that public health protections in premarket and postmarket assessments depend upon how they are *administered* and the science utilized by agencies.

Toxic contamination is unavoidable; there is no place for the public to hide. Excellent premarket testing under The Lautenberg Act *could primordially* prevent many risks of diseases that frustrate a lifetime arc of good health. However, this does not seem to be a priority for the current presidential administration, and, they have admitted as much, seeking to take regulatory burdens off the backs of private enterprise (Gibbons 2019). Removing existing toxicants will likely continue to be slothful, despite a law with "rigorous deadlines" for removing existing toxicants from commerce. They likely will linger for decades, some "forever," such as C8.

Collectively we can do much better. We need laws that primordially prevent risks of chronic and other diseases to the public from toxic exposures and undermine their lifelong opportunities would come with good health. Scientists also need to continue to develop tools, as they have done, to more quickly identify toxicants that undermine a lifetime arc of good health.

However, finally we need presidential administrations and environmental health administrators committed to using scientific defaults and the best science for protecting the public's health.

References

Armstrong GL, Cohn LA, Pinner RW (1999) Trends in infectious disease mortality in the United States during the 20th century. JAMA 281:61–66

Cao J, Xu X, Hylkema MN (2016) Early-life exposure to widespread environmental toxicants and health risk: a focus on the immune and respiratory systems. Ann Global 82:119–131. https://doi.org/10.1016/j.aogh.2016.01.023

Cogliano VJ, Baan RA, Straif K et al (2008) Use of mechanistic data in IARC evaluations. Environ Mol Mutagen 49:100–109

Cohn BA, Wolff MS, Cirillo PM, Sholtz RI et al (2007) DDT and breast cancer in young women: new data on the significance of age at exposure. Environ Health Perspect 115(10):1406–1414

Corrosion Proof Fittings v. EPA, 947 F.2d 1201 (5th Cir. 1991)

Cranor CF (2011) Legally poisoned: how the law puts us at risk from toxicants. Harvard University Press, Boston

Cranor CF (2017a) Tragic failures: how and why we are harmed by toxic substances. Oxford University Press, New York and Oxford

Cranor CF (2017b) EPA scientists said ban the pesticide chlorpyrifos. Scott Pruitt said no. Los Angeles Times, located at http://www.latimes.com/opinion/op-ed/la-oe-cranor-chlorpyrifos-should-be-banned-20170606-story.html, June 7, 2017

Cranor CF (2018) "Proof, presumptions and defaults," presented at "Understanding Pathways to a Paradigm Shift in Toxicity Testing and Decision-Making," Proceedings of a National Academy of Sciences Workshop, November 20–22, 2017

Daniels N (1981) Health-care needs and distributive justice. Philos Public Aff 10(2):146–179

Findwords. Available at https://findwords.info/term/bodily%20integrity

Frank R. Lautenberg Chemical Safety for the 21st Century Act, Public Law No: 114-182, signed into law, June 22, 2016

Fries JF (1980) Aging, natural death, and the compression of morbidity. N Engl J Med 303(3):130–135

Fries JF, Bruce B, Chakravarty E (2011) Compression of morbidity 1980–2011: a focused review of paradigms and progress. J Aging Res, Article ID 261702

Gibbons S (2019) 15 ways the Trump administration has changed environmental policies. National Geographic, located at https://www.nationalgeographic.com/environment/2019/02/15-ways-trump-administration-impacted-environment/

Grandjean P, Bellinger D, Bergman A et al (2008) The Faroes statement: human health effects of developmental exposure to chemicals in our environment. Basic Clin Pharmacol Toxicol 102:73–75

Heath CW et al (1975) Characteristics of angiosarcoma of the liver among PVC workers in the U.S. in special issue: toxicity of vinyl chloride-polyvinyl chloride. Ann N Y Acad Sci 246:231–236

Heindel J (2008) Animal models for probing the developmental basis of disease and dysfunction paradigm. Basic Clin Pharmacol Toxicol 102:76–81

Heindel J (2018) The developmental basis of disease: update on environmental exposures and animal models. Basic Clin Pharmacol Toxicol:1–9. https://doi.org/10.1111/bcpt.13118

Huen K, Harley K, Brooks J, Hubbard A, Bradman A, Eskenazi B, Holland N (2009) Developmental changes in PON1 enzyme activity in young children and effects of PON1 polymorphisms. Environ Health Perspect 117(10):1632–1638

Julvez J, Smith GD, Golding J, Ring S, St. Pourcain B, Gonzalez JR, Grandjean P (2013) Prenatal methylmercury exposure and genetic predisposition to cognitive deficit at age 8 years. Epidemiology 24(5):643–650

Lerner S (2015) The teflon toxin: the case against DuPont. The Intercept. Available at https://theintercept.com/2015/08/17/teflon-toxin-case-against-dupont/

Lombardi K (2014) Benzene and worker cancers: an American tragedy. Center for Public Integrity. Available at http://www.publicintegrity.org/2014/12/04/16320/benzene-and-worker-cancers-american-tragedy

Mancini J (2017) DuPont reaches C8 Settlement agreement for $670M. The Parkersburg News and Sentinel. Available at http://www.newsandsentinel.com/news/local-news/2017/02/dupont-reaches-c8-settlement-agreement-for-670m/

Mayo Clinic, Infectious diseases, located at https://www.mayoclinic.org/diseases-conditions/infectious-diseases/symptoms-causes/syc-20351173

MedicineNet; Medical Definition of Chronic disease, located at http://www.medicinenet.com/script/main/art.asp?articlekey=33490

Milward v. Acuity Specialty Products, Inc., 639 F.3d 11 (2011)

Morris J (2015a) 'A toxic legacy', Toxic substances in electronics manufacturing: the U.S. does tragically little to protect workers from them. Money Box: Commentary about Business and Finance. Available at http://www.slate.com/articles/business/moneybox/2015/07/toxic_subst…s_in_electronics_manufacturing_the_u_s_does_tragically_little.html

Morris J (2015b) She loved making people feel great: Sandy Guest, 55, hairdresser. Center for Public Integrity. Available at https://publicintegrity.org/workers-rights/read-their-stories-how-job-related-illnesses-upended-these-families-lives/

National Research Council (NRC) (2009) Science and decisions: advancing risk assessment. National Academies Press, Washington, D.C. https://doi.org/10.17226/12209

Perera FP, Jedrychowski W, Rauh V, Whyatt RM (1999) Molecular epidemiologic research on the effects of environmental pollutants on the fetus. Environ Health Perspect 107(S3):451–460

Rawls J (1999) A theory of justice, revised edition. Harvard University Press, Cambridge

Schardein JL (2000) Chemically induced birth defects, 3rd edn., rev. and expanded. Marcel Dekker, New York

Smith MT, Guyton KZ, Gibbons CF et al (2016) Key characteristics of carcinogens as a basis for organizing data on mechanisms of carcinogenesis. Environ Health Perspect. https://doi.org/10.1289/ehp.1509912

Trasande L, Liu Y (2011) Reducing the staggering costs of environmental disease in children, estimated at $76.6 billion in 2008. Health Aff 30:1–8

US Congress, Frank R. Lautenberg Chemical Safety for the 21st Century Act, Public Law No: 114-182, signed into law, June 22, 2016

US Congress, Office of Technology Assessment (1987) Identifying and regulating carcinogens. US Government Printing Office, Washington, DC

US Council on Environmental Quality (1971) Toxic substances. US Government Printing Office, Washington, DC
US Department of Health and Human Services, Centers for Disease Control and Prevention (CDC) (2019) National Report on Human Exposure to Environmental Chemicals (update of Fourth annual Report). Available at https://www.cdc.gov/exposurereport/index.html
US Environmental Protection Agency (2016) Assessing and managing chemicals under TSCA: highlights of key provisions in the Frank R. Lautenberg Chemical Safety for the 21st Century Act. 2016. Available at https://www.epa.gov/assessing-and-managing-chemicals-under-tsca/highlights-key-provisions-frank-r-lautenberg-chemical
US Government Accountability Office (GAO) (2008) Chemical assessments: low productivity and new interagency review process limit the usefulness and credibility of EPA's risk integrated information system GAO-08-440. US General Accounting Office, Washington, DC (March 2008)
Vineis P, Stringhini S, Porta M (2014) The environmental roots of non-communicable diseases (NCDs) and the epigenetic impacts of globalization. Environ Res 133:424–430
Wick P, Malek A, Manser P, Meili P, Maeder-Althaus X et al (2010) Barrier capacity of human placenta for nanosized materials. Environ Health Perspect 118(3):432–436
Woodruff TJ, Zota AR, Schwartz JM (2011) Environmental chemicals in pregnant women in the United States: NHANES 2003–2004. Environ Health Perspect 119(6):878–885

Epidemiology and Public Health Under Siege: In Whose Best Interests?

Colin L. Soskolne

Abstract Epidemiology is a core science that informs the development of rational public health policy. The obligation of epidemiologists, as part of the community of health scientists, is to protect public health and safety from established, as well as suspected avoidable harms. This obligation is discussed in the context of ecological and related environmental damage. It underscores why ecological integrity is a concern for those in public health and, indeed, with the survival of our species.

Because health protection is central to public health's mission, the four-level framework for the prevention of disease and premature death is considered; namely: Primordial, Primary, Secondary, and Tertiary Prevention. The first two of these are more related to concerns for ecological integrity, tying nicely into preventing our species' demise and possible extinction.

Being an applied, soft science, epidemiology provides the core scientific methods for health protection. Epidemiological evidence—coming from a soft science—can be influenced by interests that manipulate the scientific method to cast doubt, thus fomenting uncertainty designed to mislead both the public and policy-makers. Because of this, some ethical and practical dimensions of influence that result in delayed policy action are identified, to help us better control influences that derail science counter to the public interest.

Some examples highlight the challenges posed, making the solutions that follow more tractable. To protect the public interest and better ensure the sustainability of life on Earth, the need for vigilance by professionals with the capacity to speak truth to power is noted. Additionally emphasized is the need to train students to diagnose scientific misconduct and to devise and provide effective strategies for calling it to account.

C. L. Soskolne (✉)
University of Alberta, Edmonton, AB, Canada
e-mail: colin.soskolne@ualberta.ca

L. Westra et al. (eds.), *Ecological Integrity in Science and Law*,
https://doi.org/10.1007/978-3-030-46259-8_7

1 Introduction

How do we know if a disease is caused by exposure to a particular substance? And, what evidence can be relied upon to support such a connection? Who, if anyone, should care?

To answer such questions, the discipline of epidemiology bridges two domains: that of toxicology through its focus on studying disease causation in animal models examined in controlled laboratory settings, and the experience seen in human populations in the real world. Epidemiology is thus the applied, public-interest science relied on for informing rational policy formulation to protect public health and safety. As epidemiologists, we have a duty of care in such matters.

Epidemiology best informs policy through its systematic conduct and evaluation of research on human populations. It is this evidence that guides individuals, constituting the public, concerning the most precautionary measures that they might take to prevent illness and premature death. If we accept that one role of government is to protect public health, then this is achieved by establishing regulatory regimes that incentivize good conduct on the part of polluting industries, as well as best behaviours among members of the public exposed to such sources of exposure, and by instituting disincentives for poor conduct. The goal is to avoid exposures to potentially harmful products that, for reasons beyond the scope of this chapter, enter commercial markets locally, nationally and globally (Cranor 2011).

From this, it is apparent that one dimension of avoiding harm focuses on individual conduct, assuming that personal choice is available to avoid exposure with its consequent elevation of risk of harm. The other focus is on what it takes for governments to ensure an environment that is free of harmful pollutants. In the one instance, the individual has personal control over his/her exposure. On the other hand, some exposures are pervasive in the environment over which the individual has little to no control as to whether or not he/she will be exposed. Some exposures also are addictive (like tobacco products, alcoholic beverages, and a variety of drugs).

It is the obligation of professionals in public health not only to care about such matters, but also to use appropriate scientific methods to draw valid conclusions for use in informing rational policy; that is, policy based on evidence rather than policy that is either ideologically or non-fact driven, or driven on the basis of invalid science. In summary, the primary interest of public health researchers and practitioners while serving as public health scientists is to care about the public interest above any other interest.

Reports based on poor science, or misleading reports from special interest groups (ideologically and/or financially driven), can foment uncertainty, confuse the public and policy-makers, and lead to delayed or damaging policies that negatively impact people (Ruff 2015, 2017) and the living systems on which they depend. While science advances through an iterative process of falsifiability and refutation (Popper 1963; Maclure 1985), this process has been corrupted by special interests that run counter to the mission of science (Michaels 2008, 2020).

Given the many conflicting interests involved in population and community health research, we must not be naïve about the forces at play that influence both science and policy. In professional practice and in our training programs, great vigilance and personal integrity are required to counter the influence of financially interested parties and corrupt/morally bankrupt governments (Soskolne 2017; Soskolne and Baur 2019).

In particular, the seduction by moneyed interests in using academics to downplay or deny the seriousness of hazards must be recognized. These are the studies that will infiltrate the scientific literature to cast doubt and foment uncertainty (Michaels 2008, 2020). So infamous are some examples of these misdeeds that books and documentaries have explored and exposed them. The epidemiologist made aware of these misdeeds will be better equipped to recognize their appearance in other contexts.

2 Precaution and the Precautionary Principle

Sometimes, especially in dealing with complex mixtures of exposure, it is not possible to generate compelling evidence, one way or the other, in terms of identifying that substance which should be regulated to eliminate exposures. When we are concerned about environmental exposures to which large-scale population exposures occur, then, even in the presence of scientific uncertainty, limited evidence supports action. (Hill 1965; Soskolne 2005). It is the Precautionary Principle (PP) that encourages policies to protect human health and the environment in the face of uncertain risks (Kriebel et al. 2001).

In 1965, Hill provided what has become the foundation for rational approaches to using science to inform policy. Because uncertainty is inherent to all scientific knowledge, those in the applied science of epidemiology have embraced Hill's concluding remarks in which he posits:

> All scientific work is incomplete – whether it be observational or experimental. All scientific work is liable to be upset or modified by advancing knowledge. That does not confer upon us a freedom to ignore the knowledge we already have, or to postpone the action that it appears to demand at a given time (p. 300).

The need for action is required in the face of knowledge. The presence of uncertainty shall not be used as a reason for postponing cost-effective measures to prevent potentially harmful exposures, especially when they are pervasive. The role of the PP is to provide transparency in weighing risks from action (i.e., a policy intervention) against risks from inaction.

The need to be preventative and invoke the PP when harms could be so far-reaching globally becomes all the more significant and deserving of attention (Soskolne 2005).

3 Different Levels of Prevention in Public Health

Epidemiology training until the early 1990s included only three levels of prevention when discussing chronic, non-communicable diseases. These were: Primary, Secondary and Tertiary Prevention. It was in 1993, and then through its second edition in 2006, that the World Health Organization (WHO) released the textbook *Basic Epidemiology* (Bonita et al. 2006) in which a fourth level (Primordial Prevention) was included. Today, all four levels of prevention comprise the Prevention Framework. In summary, these are:

Primordial Prevention This was stimulated by the recognition of upstream determinants of health and safety. It is considered that prevention could be maximally addressed if the source of exposure could be eliminated, preventing penetration of the risk factor into the population. Thus, the need to focus efforts on removing even the possibility of exposure (for instance, eradicating tobacco plants and thus eliminating the source of tobacco exposure and hence the need for Primary Prevention).

Primary Prevention To reduce the risk of illness and premature death, public health professionals use health promoting messages to motivate people to do all in their power to avoid continued exposure to any substance/behaviour that might be harming their health and well-being. There are two strategies that are invoked here: the first is the Population Strategy; the second is the High-Risk Individual Strategy.

Secondary Prevention Where people have been exposed, Secondary Prevention is invoked to screen those who have been exposed to determine if clinical and other changes may be apparent. Early detection drives this approach because the earlier that the medical system intervenes in the progression of illness, the more likely it will be that a person can be spared from declines in health and reduced life expectancy.

Tertiary Prevention Where people who were exposed have become ill, Tertiary Prevention applies to rehabilitation efforts that might restore some level of normal function resulting in a reasonable ongoing quality of life.

It is the first two of these Levels of Prevention that are of greatest potential in the realm of ecological declines. The latter two can be helpful in the work of those doing cost-benefit analysis for determining costs saved through the implementation of effective Primordial and Primary Prevention strategies.

4 The Four Ds

When competent and impartial applied health scientists studying diseases where they arise discover a finding that does not support the status quo, going contrary to the interests of a powerful stakeholder, the epidemiologist must be prepared to face the "Four D's." These are applied with the intent of avoiding responsibility and culpability from harms caused.

The scientist can be confronted with:

- Deny—denial that the findings could be correct
- Delay—in that more research will be called for
- Divide—in that commissioned work will result in biased findings
- Discredit—if the scientist persists, he/she will be discredited.

This paradigm (i.e. the "Four D's") was applied many times over in the case of each of the following harmful substances before public policy was changed:

- Tobacco
- Nickel
- Benzene
- Lead
- Asbestos
- Climate Change

and continues today with updates provided, for example, on:

- Asbestos (Miller 2019)
- Cell phone Electro-magnetic Frequency (EMF)/Radio-Frequency Radiation exposures in cell phones (Sasco 2018; Miller et al. 2019)
- Mercury (Mergler 2019)

5 How Manipulation and Misconduct Operate

To effect delays noted above, and to drive division among scientists, epidemiologists knowledgeable about the tools of epidemiology are capable of introducing bias in subtle and influential ways into research studies. For those epidemiologists who wittingly (and usually for large sums of money) allow themselves to cast aside their scientific values and their duty to pursue truth in the public interest, a toolkit of techniques is available to skew results and produce junk science (Cranor 2011; Soskolne 2017). They seek to achieve this through applying the following techniques that usually operate at the level of applying the methods of the discipline:

- Under-powered studies
- Inadequate follow-up methods
- Inadequate follow-up time
- Inappropriate biomarkers of exposure
- Contaminated controls
- Unbalanced discussion
- Selective disclosure of competing interests
- Biased/selective interpretation
- Mechanistic information is ignored for inferring effects
- Exaggerated differences are made between human and toxicology studies, the insistence being on separating effects seen in animals from effects in humans
- Molecular structures predicting hazard potential are ignored
- The insistence on first demonstrating effects in local populations of exposed people, despite demonstrated effects in humans elsewhere
- The failure to make explicit the implicit value judgements that go into deciding appropriate standards of evidence for drawing policy-relevant conclusions (i.e., suppressing dominant interests and values).

These techniques are more fully explicated in Cranor (2011) and in Soskolne (2017), and also in Baur et al. (2015, 2019), Soskolne and Baur (2019), and Soskolne (2015, 2019). Malfeasance applies when such methods are used to support vested interests in their wish to downplay their culpability by denying, delaying, creating division and discrediting scientists.

6 Conclusions and Recommendations

The need for vigilance by professionals is called for, ideally through professional organizations with the capacity to speak truth to power. In addition, the need is ever present for the training of students capable of diagnosing scientific misconduct and with the support to call it to account. Readers ought to be aware that the International Society for Environmental Epidemiology (ISEE 2001) offers a procedure for providing moral support to beleaguered colleagues being subjected to the "Four D's."

References

Baur X, Budnik LT, Ruff K, Egilman DS, Lemen RA, Soskolne CL (2015) Ethics, morality, and conflicting interests: how questionable professional integrity in some scientists supports global corporate influence in public health. Int J Occup Environ Health 21:172–175

Baur X, Soskolne CL, Bero LA (2019) Commentary. How can the integrity of occupational and environmental health research be maintained in the presence of conflicting interests? Environ Health 18:93. Published on: 4 November 2019. https://ehjournal.biomedcentral.com/articles/10.1186/s12940-019-0527-x (accessed May 9, 2020)

Bonita R, Beaglehole R, Kjellstrom T (2006) Basic epidemiology, 2nd edn. World Health Organization, Geneva. https://books.google.ca/books/about/Basic_Epidemiology.html?id=AAZGobMNTXgC&printsec=frontcover&source=kp_read_button&redir_esc=y#v=onepage&q&f=false (accessed May 9, 2020)

Cranor CF (2011) Legally poisoned: how the law puts us at risk from toxicants. Harvard University Press, Boston

Hill AB (1965) The environment and disease: association or causation? Proc R Soc Med 58:295–300

International Society for Environmental Epidemiology (2001, revised 2009) ISEE Procedure for Dealing with Beleaguered Colleagues and/or Potential Whistleblowers. https://www.iseepi.org/Public/Public/About_Us/ISEE_Committees/Ethics_and_Philosophy.aspx. (accessed May 9, 2020)

Kriebel D, Tickner J, Epstein P, Lemons J, Levins R, Loechler EL, Quinn M, Rudel R, Schettler T, Stoto M (2001) The precautionary principle in environmental science. Environ Health Perspect 109(9):871–876. https://www.ncbi.nlm.nih.gov/pubmed/11673114 (accessed May 9, 2020)

Maclure M (1985) Popperian refutation in epidemiology. Am J Epidemiol 121:343–350

Mergler D (2019) The legacy of mercury poisoning and racism: the case of a First Nation community in Canada. Abstract. Annual Ramazzini Days, 24–27 October; Carpi, Italy

Michaels D (2008) Doubt is their product: how industry's assault on science threatens your health. Oxford University Press, New York

Michaels D (2020) The Triumph of Doubt: Dark Money and the Science of Deception. Oxford University Press, New York, USA

Miller A (2019) 20 years after the Libby Montana asbestos response: past, current, and future issues and implications. Abstract. Annual Ramazzini Days, 24–27 October; Carpi, Italy
Miller AB, Sears ME, Morgan LL, Davis DL, Hardell L, Oremus M, Soskolne CL (2019) Risks to health and well-being from radio-frequency radiation emitted by cell phones and other wireless devices. Front Public Health 7:223; 10 pages. https://doi.org/10.3389/fpubh.2019.00223.; https://www.frontiersin.org/articles/10.3389/fpubh.2019.00223/full. Accessed 9 May 2020
Popper K (1963) Conjectures and refutations: the growth of scientific knowledge. Routledge, London
Ruff K (2015) Commentary: scientific journals and conflict of interest disclosure: what progress has been made? Environ Health 14:45. https://doi.org/10.1186/s12940-015-0035-6. (accessed May 9, 2020)
Ruff K (2017) Serving industry, promoting skepticism, discrediting epidemiology, Chapter 7. In: Walker MJ (ed) Corporate ties that bind: an examination of corporate manipulation and vested interest in public health. Skyhorse Publishing, New York, pp 119–135; 482–485
Sasco AJ (2018) Has anything changed since the 2011 IARC classification of EMF? Abstract. Annual Ramazzini Days, 1–4 November; Carpi, Italy
Soskolne CL (2005) On the even greater need for precaution under global change (Reprinted from The European Journal of Oncology Library, Vol 2, 2003). Human Ecol Risk Assess 11:97–106
Soskolne CL (2015) Public health and environmental health risk assessment: which paradigm and in whose best interests? In: Westra L, Gray J, Karageorgou V (eds) Ecological systems integrity: governance, law and human rights. Earthscan, London, Chapter 16, pp 191–200
Soskolne CL (2017) Global, regional and local ecological change: ethical aspects of public health research and practice. Part 1, Chapter 1. In: Zölzer F, Meskens G (eds) Ethics of environmental health. Earthscan from Routledge, Oxon. [ISBN 978-1-138-18662-0 (hbk); 978-1-315-64372-4 (ebk)]; pp 3–16
Soskolne CL (2019) The role of vested interests and dominant narratives in science, risk management and risk communication. Chapter 8. In: Zölzer F, Meskens G (eds) Environmental health risks: ethical aspects, Routledge studies in environment and health. Routledge, Taylor & Francis. London and New York, pp 123–134
Soskolne C, Baur X (2019) How corporate influence continues to undermine the public's health. Commentary. J Sci Pract Integr 1(1). file:///C:/Users/15142/Downloads/9747-how-corporate-influence-continues-toundermine-the-public-s-health%20(2).pdf

Saving Species, Healthy Humanity: The Key Role of Women in Ecological Integrity

Shauna M. Lange

Abstract What is it that uniquely qualifies women to achieve success in the defense of human rights and the environment? Are there certain attributes or experiences that lead to successful activism? Societal norms and physiological predispositions have demanded certain obedience and/or nurturing roles that have led to the development of a dedication to open communication and collaboration. It is in that realm that we understand why big business and big government fear a woman with a voice.

1 Introduction

What is it that makes women uniquely in-tune to their environment? What does history tell us about the changes made by women in ecological integrity? Ultimately, what can be done to protect and encourage this type of involvement in environmental conflict and advocacy?

Jane Goodall, Marjory Stoneman Douglas, Greta Thunberg, Berta Isabel Caceres Flores and our very own Laura Westra; these women, these awe-inspiring warriors, are just a few who have not only enhanced the world view of ecological integrity but also helped to more clearly define it. Whether advocating for animals, protecting natural resources, or seeking punishment and accountability for environmental harms, these dedicated she-soldiers have championed their causes resulting in dramatic change.

Environmental Justice is the fair treatment and meaningful involvement of all people regardless of race, color, national origin, or income with respect to the development, implementation, and enforcement of environmental laws, regulations, and policies. For centuries women have fought for equal rights, fair treatment, and the right to be heard. In today's world, these battles have prepared them to fight the

S. M. Lange (✉)
Rees and Lange, P.C., Hobart, IN, USA
e-mail: Shaun@reeslegal.com

L. Westra et al. (eds.), *Ecological Integrity in Science and Law*,
https://doi.org/10.1007/978-3-030-46259-8_8

good fight, namely, seeking justice for those (flora or fauna) who cannot seek it for themselves.

To fully understand, we must investigate the skills, attributes, and societal norms that have both encouraged and obstructed women's ability to cause meaningful change and participate productively in a predominantly male-driven world. Our analysis will help us gain insight into what makes women so well-equipped to flourish at the forefront of environmental activism, what challenges they face, and how we can achieve a more secure platform where others feel inspired to join the cause.

2 The Who's Who?: Celebrating Accomplishments

As environmental activism becomes more and more dangerous, we often reflect upon those who came before us or those from whom we take inspiration. This is by no means an exhaustive list, but what follows in this section is an overview of several women environmental defenders who have utilized their unique feminine expertise to truly make a difference.

2.1 *Dr. Jane Goodall (1934)*

We begin our venture into the lives of these incredible ladies with a household name, Dr. Jane Goodall. Huffington Post reported that Dr. Goodall "has transformed the epistemological framework of how we study primates, and has exemplified immeasurable ecological integrity" (Pierrat 2017). For nearly sixty years, Dr. Goodall has been the world's foremost expert in the study of ecology and chimpanzees.

So, we ask, what sets her apart? Dr. Goodall defied the norms of scientific study by "harnessing the power of the feminine" to validate her work (Pierrat 2017). Instead of focusing on chimpanzees as another number or thing to be categorized abstractly, she gave them names, noted their emotions, their personalities, the behaviors that make each one distinctly unique (Pierrat 2017). By doing so, she brought ecological integrity at its finest into the work of herself and her colleagues, forever changing the way we study animals.

2.2 *Marjory Stoneman Douglas (1890–1998)*

Our next SHEro is Marjory Stoneman Douglas, this brilliant journalist turned environmental pioneer discovered that not only were The Everglades in the United States a vital part of the American ecosystem, but they were in fact a "vibrant web of ecosystems" tied and thriving together (Handcock 2019). She recognized the importance of this one of a kind source of fresh water and changed its treatment forever. As the women's rights movement was taking on new challenges in the United States, she was pushing the limits penning books to influence popular opinion in another way. The late Florida (USA) Governor Chiles expressed his support for her work, calling Ms. Douglas " a prophet, calling out to us to save the environment for our children and our grandchildren" (Handcock 2019). Again, humanizing the issues with a woman's touch.

2.3 *Greta Thunberg (2003)*

Current events have no doubt introduced everyone to a young Swedish female activist named Greta Thunberg. This teenager led the largest climate strike in history, which was said to include roughly four million people from more than one hundred forty-one countries around the world. A look at her counterparts around the world, those children and adults who support and stand with her, reveals that an overwhelming majority of them are women.

Specific to Ms. Thunberg on a personal level, she has been congratulated for her refusal to "kowtow" to the patriarchy (Moore 2019). Political leaders, government officials, and influential business men have criticized Thunberg not based upon her actions, but for everything from her looks and her demeanor, to the tone of her voice and her clothes. The Guardian reported:

> The extraordinary reactions of certain men to Greta Thunberg, which have nothing to do with her urgent and necessary message on the climate emergency, tells you exactly what happens when a young woman simply refuses to be sexualized. It makes them deeply uncomfortable (Moore 2019).

Ms. Thunberg is showing the world firsthand what it means to be a strong woman in the environmental movement and how to shut down critics by refusing to fit into the stereotypical fantasy of the global patriarchy.

2.4 *Berta Isabel Caceres Flores (1971–2016)*

The tragic death of this environmental heroine was no accident. Ms. Flores was assassinated for her work to protect the lands of her people. Ms. Flores brought attention and legal action to a situation where the developers had "broken international law by not consulting with local tribes, who were concerned that the dam would prevent Lenca communities from accessing water, food, and materials for medicine" (Handcock 2019).

Present throughout Ms. Flores' life and fight to preserve the natural world is an overarching theme of protecting the vulnerable, providing for others and "fighting for a life with dignity" (Handcock 2019) She was known as "La Guardiana," the Guardian, protecting Mother Earth from those who would harm her, like a mother protecting her child.

2.5 *Dr. Laura Westra (1933)*

Our final championess of ecological integrity is our very own Dr. Laura Westra. For more than 27 years, Dr. Westra has organized and led the Global Ecological Integrity Group, a group of more than 250 scholars, lawyers, doctors and independent researchers from a variety of disciplines who come together to "push the boundaries of scholarly endeavour through inter- and trans-disciplinary engagement on matters affecting and governing the sustainability of life for both present and future generations" (Westra 2019).

Aside from her work with the Global Ecological Integrity Group, Dr. Westra is a Professor, an author with more than one hundred titles to her name, and an enthusiastic speaker regarding issues concerning environmental policy and ethics. She also serves on the board for many professional organizations.

Like the women above, Dr. Westra is passionate about the intersection between ethics and our environment. She advocates tirelessly through her teachings, presentations and publications for ecological integrity and the responsibility of humans as a global society. Her work fosters communication, inspires responsible action and encourages collaboration, all results that are characteristically tied to women-led initiatives.

Besides the similarities in their chosen passions, the aforementioned women illustrate for us a pattern that clearly emerges when women apply themselves in the ecological integrity arena. Whether these women chose their path or unexpectedly found themselves in the middle of it, their success provides us with an understanding of what attributes make women uniquely suited for the job.

3 Whether Willing or Unwitting, Why Do They Act and What Sets Them Apart?

Historically, studies of environmental conflict have shown that "because of their close contact with the environment and use of natural resources, women are often affected by environmental conflicts" (Asteria et al. 2013) on a much larger scale than men.

3.1 *Recognizing Skills and Characteristics*

Women continue to demonstrate via their movements and actions that they have an immense capacity to serve as educators, leaders, and experts when it comes to protecting the environment, resources, and communities around them.

Global Fund for Women reported in their 2019 Earth Day piece that:

> Protecting land and natural resources from climate change, increasing state corruption, megaprojects spearheaded by powerful corporations, and systemic armed violence have

pushed indigenous women to mobilize their communities and lead the way to protect their survival and way of life (Global Fund for Women 2019).

Whether they chose the path of an activist or are reactionary to the aforementioned conflicts as they arise around them, certain characteristics and skills that are inherent in women positively affect their ability to successfully facilitate change.

3.2 The Indonesian Studies and the Skills of Communication, Empathy and Dedication to Quality of Life

In a study of three urban areas in Indonesia suffering from environmental conflict, research showed that some of these characteristics are "self-orientation to quality of life, care, gentleness, empathy, and advanced aspects of parenting" (Asteria et al. 2013). These Indonesian studies, consisting of 140 questions distributed to activists in the form of a questionnaire, were conducted in areas dealing with the scarcity of clean water, waste, and pollution due to domestic industry and/or river pollution (which is a source of clean water) because of a slaughterhouse (Asteria et al. 2013).

It is said that "Environmental management and maintenance require empathy in human interaction with the environment" (Asteria et al. 2013). Accordingly, it appears that attributes that many have considered to portray women as the "gentler sex" can be the very attributes that empower them to act during environmental crisis.

Present throughout the same female environmental activism, whether intentional or not is the concept of eco-feminism. Eco-feminism, focused on "examining the connections between women and nature" (Miles 2013) ecofeminism adds both a commitment to the environment and an awareness of the associations made between women and nature. Specifically, this philosophy emphasizes the ways both nature and women are treated by society.

It has been argued that, "women tend to have more concern for the environment because of its nurturing character" (Asteria et al. 2013). As such, when women activists seek to solve an environmental crisis, they have been shown to utilize strong communication and mediation 'to build mutual understanding and avoid hostility between the two sides of the conflict" (Asteria et al. 2013). In doing so, they take proactive action which may avoid exacerbating the tensions.

The Indonesian studies found that overall, "women activists prioritize communication goals made in order to keep harmonious relationship. Communication styles of women activists who do have a tendency to be cooperative because women have an orientation maintaining relationships in the future and for the overall welfare of the community" (Asteria et al. 2013). These efforts in conflict resolution are known as "musyawarah…a process of shared decision-making that the goal is to reach a consensus" (Asteria et al. 2013).

Following this open communication in which women tend to foster a resolution-seeking mentality for all parties, it has been found that the result is "empowering the community" (Asteria et al. 2013). When an open communication goal is achieved, the community will retain their autonomy while being able to actively engage in mediation, negotiation or any other form of dialogue in which they work towards a resolution.

3.3 Gender Justice, Environmental Justice and Gender Equality

Along with the need for open dialogue and communication, women recognize the importance of gender justice and gender equality (not to say that men do not) as an integral part of ecological integrity and environmental action. In the pursuit of gender equality, gender justice is the main foundation. It is imperative to women in this capacity that they ensure that the process is fair for women and men to take action to stop things that socially and historically discourage either sex from participating.

Environmental Justice is the fair treatment and meaningful involvement of all people regardless of race, color, national origin, or income with respect to the development, implementation, and enforcement of environmental laws, regulations, and policies. Women tend to open the lines of communication and foster an environment where everyone works towards a common goal as opposed to engaging in a battle of wills. Given the struggles throughout history of woman seeking equality in every day life, they often find themselves in a position of responsibility and/or obligation. In speaking of her activism to protect indigenous lands from a hydroelectric dam project before her assassination in 2016, Berta Cáceres said, "Mother Nature—militarized, fenced-in, poisoned—demands that we take action" (Global Fund for Women 2019).

Throughout history, women have overcome so many challenges in seeking equality and the fight continues. Experience in addressing issues of inequality and failed justice leads to later success stories and foundational principals for getting things done. Just as there are many different roles, functions, and responsibilities between men and women formed from socio-cultural constructions, the same variety exists in our goals, and varied ability to facilitate change (Nafi et al. 2018). As a society, our desire to cling to these historical norms and constructions hinders our efforts towards change but there is always hope.

3.4 Women Are Global Thinkers

Another skill possessed by so many woman is the consideration of global thinking. Dr. Jane Goodall once said that "One individual cannot possibly make a difference, alone. It is individual efforts, collectively, that makes a noticeable difference—all the difference in the world!" (Ashoka 2013) This is yet another reason why woman foster open communication and mediation in environmental conflict.

Woman in environmental activist roles often see the impact their work will have not just for their immediate location or community, but on a broader scale. Miriam Miranda, a leader of the Black Fraternal Organization of Honduras (OFRANEH), which defends indigenous land rights and natural resources said: "[I fight] so that future generations will have the resources and assets they need to survive on a planet that's being destroyed every day" (Global Fund for Women 2019). However, there are some roles, stereotypes and norms that both hinder and empower women during environmental conflict.

4 Traditional, Cultural and Societal Roles as Barriers and Empowerment

Some of our greatest assets and skills cause our greatest detriment. For example, we have seen that:

> the struggles of women climate activists are gaining more traction and visibility as attacks and violence against them continues to grow in response to their relentless activism—but their voices are still largely missing in global conversations about climate change, sustainability, and access to land, water, and natural resources (Global Fund for Women 2019).

The barriers of gender inequality and a lack of involvement in the global arena pose distinct challenges when it comes to battling cultural and societal norms. The way society views the role of women has created another hurdle to jump before women can actually focus on the issues surrounding the environmental conflict that affects them. Inspiring still, we are seeing an increase in participation in environmental management and the global and local political arena.

4.1 Women as Providers

In many cultures, women play key roles in defending the environment. They face increased stress and insurmountable workloads when displacement or separation from resources occur (Nafi et al. 2018). This is because it is often their job to conserve biodiversity, and provide sustenance to their communities by collecting and harvesting food and medicines (Nafi et al. 2018). With this great responsibility comes also the burden of ensuring success as their connection to the environment and their duties are a matter of life and death.

4.2 Women Are Nature

In other cultures, women are seen as the leaders of the communities, the mother, the grandmother, the nurturer tied to nature, and as such, the one who must protect it as her own. This lends itself to community support but also to very challenging expectations. Women can be considered a part of nature, as nature itself, often "Mother Nature" is considered a woman. This sometimes results in stereotypes of weakness, vulnerability, and a need for protection from the males of our species rather than a sign of expertise. However, the fact that women and their varying roles and duties are often so intertwined with nature and the world around them that they know it best of all. As such, they are the best equipped to support, fight for, and maintain it.

4.3 *Women as Caregivers*

Some realms of society choose to view women predominantly as caregivers and many women also view themselves that way. Reyna Ortiz, defender in El Salvador with grantee partner Movimiento de Mujeres de Santo Tomas supported the theory, stating "I fight because I know that without water, we cannot live. I do this work for the love of my community, for my granddaughters…so that they can live in a healthy world" (Global Fund for Women 2019). However, at some point,

> This function of care-giver becomes a value assigned by the patriarchal world. It also has its positive aspects since it is what ensures there are women caring for the water, or engaged in the environmental movement, or that the Ruta Pacifica exists- a movement of women for peace and a negotiated settlement to the armed conflict in Colombia. But it has been "too much", as if caring were only women's responsibility. What is missing is to share the caring, a more equitable distribution of caring

that equally represents those who benefit (Mendoza 2015).

The caregiver view lends itself to the limiting concept that women shall always give priority to the needs of others and in turn, that women exist for others and are only present for that purpose (Mendoza 2015). Alternatively, and as previously mentioned, acting in a caregiver capacity can make someone an expert regarding their ward and as such, they are in the best position to advocate for the same.

On the opposite side of the coin, there are some challenges that don't stem from anything other than physically being women.

5 When Stereotypes Attack Both Literally and Figuratively

Perhaps Margaret Atwood said it best when she explained "We still think that a powerful man is a born leader and a powerful woman is an anomaly" and "Men are afraid that women will laugh at them. Women are afraid that men will kill them." Some brush these statements off as flippant but there is veracity therein.

5.1 *Starting Behind the Mountain*

In environmental and human rights activism, women are targeted for who they are as women and for this reason, many must overcome the mountain of gender inequality before they are able to address the actual issues of the pending conflict. What is much worse, is the fact that old patriarchal views attempt to hold women to ancient roles. When a woman steps outside her patriarchy-designated role in the home, she may find herself chastised, criminalized, assaulted or even killed.

Jakeline Romero, a Wayuu and women's rights activist has been threatened on numerous occasions. In December of 2016, she received the following text message:

> Don't focus on what doesn't concern you [if] you want to avoid problems. Your daughters are very lovely, so stop stirring other people's pots [...] Bitch, avoid problems because even your mother could be disappeared if you keep talking (McLaughlin 2017).

Although she was continuously threatened, she has remained true to her cause against extractive industry in Colombia. These tactless and criminal attempts to silence women are found more often in the aforementioned patriarchal societies where defenders are targeted solely for their gender.

Female environmental, and territory defenders are at specific risk from gender violence generated by various assailants, including the State. Just as we noted in the verbal attacks on young Greta Thunberg, women are sexualized, marginalized and stereotyped.

> Gender-based violence is also present when women human rights defenders are subjected to sexist remarks, acts of sexual violence and threats made, against their children or families (Vidal 2018)

> These attacks are particularly problematic in community contexts; even though violence against women activists has increased in cities, the number of cases is bigger in rural communities. The most common aggressions include: false accusations and launching defamation campaigns that jeopardize women's credibility, harassment and threats. Attackers make false accusations by using activists' personal information, photoshop and slander campaigns in social media, have increased through the promotion of gossip that undermines their work (McLaughlin 2017).

Perhaps the most frightening idea here is not only are the state and local governments failing to end these attacks but the fact that in many cases, they have been said to participate in them. Governments stooping to the level of criminal involvement in defamation campaigns and much more serious attacks on these women will continue until both governments and big business are held accountable and forced to be transparent.

5.2 The Continued Failure of State and Local Governments

In the 2018 publication by the Global Ecological Integrity Group, *Ecological Integrity and Land Uses*, Chapter 10, The Silence of the State: The Failure of State and Local Governments in Protecting Environmental Defenders discussed the devastating statistics regarding the assault, murder and disappearance of environmental defenders around the world. Further study has revealed that an immense number of those attacked were women.

In the publication, "Criminalization of the work of human rights defenders" of 2015, a study showed the staggering figures and statistics affecting women just like

Jakeline Romero where instead of supporting these defenders, state and local governments engage in unwarranted prosecution and violence:

> the attacks have been aimed at defenders who are dedicated to the defense of the rights of indigenous peoples, territory, land and healthy environment. The organizations reported that 55% of people assaulted are women and 58.21% of all attacks have been aimed at environmental defenders, particularly from the perspective of indigenous peoples. Several environmental advocates are awaiting decisions in open proceedings against them (Hernández et al. 2016).

By way of further statistical explanation, the Mesoamerican Women Human Rights Defenders Initiative indicated that

> the greatest number of attacks on women human rights defenders were against those who defended land and territory. In 2012, 157 women (38% of all attacks on women human rights defenders) were attacked; in 2013, 81 women (or 15%); and in 2014, 287 women (or 38%). Alarmingly, between 2012 and 2014, 1,688 attacks on women human rights defenders were reported in just El Salvador, Guatemala, Honduras, and Mexico – with the frequency of the attacks doubling between 2012 and 2014 (DeWolfe and Maya 2017).

More specifically with respect to what methods are being used, studies are showing that:

> of those committing acts of violence against women land and territory defenders are mostly regional stakeholders: and companies that use different tactics, for example: the use of public force and criminalization; establishing criminal records by randomly and illegally detaining those participating in peaceful demonstrations or checkpoint blockages; adopting intimidation and harassment methods; making telephone threats and false accusations; launching defamation campaigns and even assassination of leaders (Vidal 2018).

Unless action is taken on a global scale, it is unlikely that these attacks will end anytime soon. The inaction and often involvement of government and public officials makes eradication of ancient viewpoints, stereotypical limitations and bogus legal action against defenders nearly impossible to stop.

We are not naïve and as such recognize that a world without environmental conflict would be impossible. However, it is not unreasonable to expect humanity to prevail and level the playing field. With an eye on the ultimate ideals, we forge on, hopeful to someday reach balance in which everyone can engage in activism without first having to battle gender stereotypes.

6 Men, Woman, Humanity….The Ultimate Goal

This chapter has shown the challenges, risks and successes of women environmental defenders. The Global Fund for Women optimistically reports that:

> Women, girls, trans people, and indigenous women land and water defenders are resilient, creative, and steadfast in their commitment to standing up for gender equality and resisting encroachment on their water and land—no matter how big the threat (Global Fund for Women 2019).

Throughout history, women have overcome societal limitations to make meaningful change. Ana Sandoval, and defender and co-founder of grantee partner Communities in Peaceful Resistance "La Puya", working to resist dangerous mining megaprojects in her community in Guatemala, said "In the end, all the struggles have the same objective: the defense of life. That is the most important, no matter where we are or what the specific goal of each fight is" (Global Fund for Women 2019).

Whether actively pursuing solutions to a relevant environmental conflict, fighting for gender equality, or another goal all together, women are uniquely qualified to defend life. Women are natural born advocates, whether it be through their personal or professional roles, they find a way. Betty Vasquez, a land defender working with grantee partner Centro de Derechos de Mujeres in Honduras, clearly explained:

> It is not an option but rather an obligation to speak out, all of us together, to demand changes. I do not imagine my life without my activism. It is my conviction to do it and makes me want to live (Global Fund for Women 2019).

As equality continues to expand to all facets of life and men and women stand together on the front lines to champion causes, the focus can shift from which roles each party must take on to how we can collectively, as a global society effectuate needed change. Ultimately, it remains the obligation of state and local governments to see this through. The specific successes and failures of governments in working on this obligation is discussed in another article but one clear example of this is present in Zimbabwe's Lower Zambeze Valley where Africa's first all women anti-poaching unit is making history:

> "There's a saying in Africa, 'If you educate a man, you educate an individual, but if you educate a woman, you educate a nation'," Mander says. "We're seeing increasing evidence that empowering women is one of the greatest forces of change in the world today" (Nuwer 2018).

In the end, state and local governments must recognize the importance of community involvement and input in major projects, foster an atmosphere of openness and accountability, and encourage participation from all citizens. After all, they are the main guarantor of security and as such, are responsible for protecting the rights of all people. Until such a time as gender equality is no longer hindered by outdated societal perception, glass ceilings, and failures of protection mechanisms, women will continue to fight both battles.

References

Ashoka (2013) 9 great quotes by women environmentalists. Forbes. Available via https://www.forbes.com/sites/ashoka/2013/04/30/9-great-quotes-by-women-environmentalists/#123f19ad10b6. Accessed 31 Oct 2019

Asteria D, Suyanti E, Utari D, Wisnu D (2013) Model of environmental communication with gender perspective in resolving environmental conflict in urban area (Study on the Role of

Women's Activist in Sustainable Environmental Conflict Management). The 4th International Conference on Sustainable Future for Human Security, SustaiN 2013
DeWolfe E, Maya JS (2017) MesoAmerican women human rights defenders initiative, violence against woman human rights defenders in Mesoamerica. In Defense of Life. Available via http://www.awid.org/sites/default/files/atoms/files/286224690-violenceagainst-whrds-in-mesoamerica-2012-2014-report. Accessed 29 Oct 2019
Global Fund for Women (2019) 8 quotes that emphasize why we should celebrate and support women land defenders for earth day. Global Fund for Women. Available via https://www.globalfundforwomen.org/8-quotes-women-land-defenders-earth-day/. Accessed 30 Oct 2019
Handcock K (2019) Guardians of the planet: 15 women environmentalists you should know. A Mighty Girl. Available via https://www.amightygirl.com/blog?p=11863. Accessed 29 Oct 2019
Hernández A, Romero P, Cerami A, Lima J, Oliveira I, Thenadey A, Persadie N, Serracin S, Lucia MB, Xiloj L, Noack J (2016) Emerging practices of states regarding the protection of environmental defenders in Latin America and the Caribbean. Mexican Center for Environmental Law AC (CEMDA). November 2016. Accessed 29 Oct 2019
McLaughlin C (2017) Defenders of the Earth. Global Killings of Land and Environmental Defenders in 2016. Global Witness
Mendoza U (2015) Echos of the sounds of the conch: calling out to the hearts of women activists, defenders of human rights and universal goods in Latin America. Fundación Cultural Javeriana de Artes Gráficas – Javegraf Bogotá. Available via https://www.universal-rights.org/wp-content/uploads/2018/01/Echoes.-UAF_LA.pdf. Accessed 31 Oct 2019
Miles K (2013) Ecofeminism: sociology and environmentalism. Encyclopedia Britannica. Available via https://www.britannica.com/topic/ecofeminism. November 14. Accessed 30 Oct 2019
Moore S (2019) Greta Thunberg's defiance upsets the patriarchy-and it's wonderful. The Guardian. Available via https://www.theguardian.com/commentisfree/2019/oct/01/greta-thunbergs-defiance-upsets-the-patriarchy-and-its-wonderful. Accessed 29 Oct 2019
Nafi TH, Lestarini R, Inayati T, Wulandhary S, Aini IN, Utari D (2018) Legal protection for women environmental activists in urban areas. E3S Web of Conferences 52, 00048 (2018). Available via https://pdfs.semanticscholar.org/99d8/e9cfd3887787fa60319b34b491f0521e77ca.pdf. Accessed 31 Oct 2019
Nuwer R (2018) Meet the brave ones: the women saving Africa's wildlife. BBC.com. Available via https://www.bbc.com/future/article/20180926-akashinga-all-women-rangers-in-africa-fighting-poaching. Accessed 31 Oct 2019
Pierrat C (2017) 8 badass environmentalists you should know, Huffington Post. Available via https://www.huffpost.com/entry/8-badass-environmentalists-you-should-know December 6. Accessed 29 Oct 2019
Vidal V (2018) Interview with Verónica Vidal, Coordinator of the Association for Women's Rights and Development (AWID). In Defense of Life. Available via https://pbi-mexico.org/news/2018-06/women-have-always-been-front-lines-cause. Accessed 31 Oct 2019
Westra L (2019) (Update). About Us. Global Ecological Integrity Group. Available via https://www.globalecointegrity.org. Accessed 29 Oct 2019

Part III
Ecological Integrity and State Obligations

African Eco-Philosophy and Its Implications for Ecological Integrity in Africa

Ngozi Finette Unuigbe

Abstract The defilement of Mother Earth by humans has caused untold damage to the environment, and is especially accelerated by commercial interests driven solely to increase profits by extracting as much as possible from Nature. Short-term human interests, fuelled by an insatiable drive to accumulate money and power, have been enshrined in various laws with total disregard for the well-being of the living Earth Community.

As a continent, Africa is laden with environmental problems that may not necessarily be peculiar to it. However, strategies to tackle them would be more efficient if they are borne out of local (indigenous) ideologies. This chapter will demonstrate how the engagement of African eco relationalism has significant potentials to preserve the ecological integrity of Mother Nature on the continent and beyond. It will begin by discussing environmental problems which are prevalent in Africa and indeed, the rest of the world; and thereafter proceed by interrogating the relationalism element of African eco philosophy and how it can be applied to address the environmental problems discussed.

1 Introduction

The global ecosystem has entered into a negative feedback loop, meaning our Earth systems are unravelling—climate change, devastating air and water pollution, vastly accelerated species extinction, and the dramatic loss of food and water systems which are necessary for sustaining all life on the planet.

The defilement of Mother Earth by humans has caused untold damage to the environment, and is especially accelerated by commercial interests driven solely to increase profits by extracting as much as possible from Nature. Short-term human interests, fuelled by an insatiable drive to accumulate money and power, have been

N. F. Unuigbe (✉)
Department of Public Law, University of Benin, Benin City, Nigeria
e-mail: ngozi.stewart-unuigbe@uniben.edu

L. Westra et al. (eds.), *Ecological Integrity in Science and Law*,
https://doi.org/10.1007/978-3-030-46259-8_9

enshrined in various laws with total disregard for the well-being of the living Earth Community. For example, most laws permit and legitimise extraction from Nature upon the simple requirement of conducting an environmental impact study and/or assessments.

The growth of capitalism spreading to every part of our planet was facilitated by the legal recognition of corporations as artificial persons capable of holding rights; whereas life-giving species and components of the Earth, such as lakes, rivers, forests and mountains have been systematically denied their inherent rights to be and to flourish.

As a continent, Africa is laden with environmental problems that may not necessarily be peculiar to it. However, strategies to tackle them would be more efficient if they are borne out of local (indigenous) ideologies. This chapter will demonstrate how the engagement of African eco relationalism has significant potentials to preserve the ecological integrity of Mother Nature on the continent and beyond. It will begin by discussing environmental problems which are prevalent in Africa and indeed, the rest of the world; and thereafter proceed by interrogating the relationalism element of African eco philosophy and how it can be applied to address the environmental problems discussed.

2 African Environmental Problems

Africa is the second largest and the second most populated continent in the world. It has 54 states and a wide variety of natural diversity. In the past years, this continent has faced many political and war issues, which made the countries poor and underdeveloped. But lately, the conditions have become better and people have started thinking of development and growth. Still, there are tribal areas where people rely on hunting and simple vegetation for food. The lack of education, poverty, and the ever-increasing population has raised many serious environmental issues in Africa. These issues have become a serious concern for not only the continent, but also for the entire world.

Africa suffers from some serious environmental problems, including deforestation, soil erosion, desertification, wetland degradation, and insect infestation. Efforts to deal with these problems, however, have been handicapped by a real failure to understand their nature and possible remedies. Conventional wisdom views the people of this region as highly irresponsible toward the environment and looks to the international community to save them from themselves. It tends to blame all of the region's environmental problems on rapid population growth and poverty. Yet, there is no conclusive evidence that Africans have been particularly oblivious to the quality of the environment, nor has the international community shown any genuine concern for it until recently. Clearly, protecting the environment of Africa is an issue that needs to be examined more carefully and incorporated into an overall strategy of sustainable economic development (Omofonmwan and Osa-Edoh 2017).

Deforestation has become one of the major factors, which is affecting the ecological balance of not only Africa, but of the entire world. The clearing of forest cover for wood and agricultural land has resulted in soil erosion, climate change, less rainfall, and many other adverse conditions. Degradation in the quality of soil and loss of fertility has also become a very serious concern. Although the government has taken initiative for growing more plants, the imbalance in nature cannot be covered by just a little effort (Penny).

Air pollution. Though the countries of this continent do not contribute much to the world air pollution, the increasing population has shown an increase in the air pollution as well. This may be due to the industries, which are set up in the parts of countries where labour is easily available. Also, there are many houses where the food is still cooked by burning wood and coal. This too, contributes to air pollution (Vidal 2016).

Water pollution is among the top problems. Africa accepts solid waste from the developed countries like America, European countries and Japan for which they get paid. But these solid wastes are not processed properly, and are dumped in the rivers and other water bodies. This results in water pollution. Human waste has also contaminated the water bodies and access to fresh and clean water has become very difficult (Fariga et al. 2018).

Energy. The major source of energy here is wood. 70% of the energy is generated by burning coal and wood, which results in deforestation, soil erosion, air pollution, and ecological imbalance. The government has started awakening people and has made many efforts to get the conditions under control. An alternate energy source will require money and the continent is not so strong financially, so, this may take a while (Avila et al. 2017).

Loss of Biodiversity. Africa has many endangered species of flora and fauna. But due to the deforestation and the need of land for agriculture, there is a huge loss of biodiversity. The uneducated and tribal people are still dependent on hunting. Many of them burn up lands and forests in order to find animals easily. This has led to the death of many other animals too (IPBES 2018).

Oil Pollution. Oil is one of the major sources of income for these countries, but this has raised many environmental issues, which are very difficult to overcome. The advanced technology used to get oil is very dangerous and hazardous for the environment. But due to the lack of education and lack of proper rules, the extraction of oil is done through the same methods. The transportation of oil through the sea route has also polluted the sea water due to improper packaging and transport (Fariga et al. 2018, pp. 50–56).

These environmental issues are not only harmful for the African continent, but also for the world's ecological balance. They are contributing to global warming which is an alarming condition. Hence, strict measures should be taken by the African government and even the world's environmental organizations to check these issues before it leads to dangerous consequences. One of such strategies—which is what this chapter argues for, is the engagement of African eco relationalism that stems from African religion, into environmental management policies and strategies. The foundation for this will be laid in the following section.

3 African Ecophilosophy: An Overview

African traditional religion is the indigenous religion of Africans whose origin is lost in antiquity. However, it is believed to have been practised by African fore bearers who handed it down from one generation to another. Awolalu has described this religion as "the indigenous religion of the Africans. It is the religion that has been handed down from generation to generation by the fore bearers of the present generation of Africans. It is not a fossil religion (a thing of the past) but a religion that Africans today have made theirs by living it and practicing it" (Awolalu 1976).

It is a religion founded on African soil, and inextricably interwoven with the culture of the people, expressed in beliefs and practices, myths and folktales, songs and dances, liturgies, rituals, proverbs, pithy sayings and names, "sacred spaces and objects; a religion which is slowly but constantly updated by each generation in the light of new experiences through the dialectical process of continuities and discontinuities" (Ekwunife 1990).

According to the African worldview, currently living human and nonhuman beings, the living dead, the yet unborn, and the natural world are interconnected (Behrens 2012). This is the basis of ecophilosophy. Drengson argues that just as the aim of traditional philosophy is Sophia (wisdom), the aim of ecophilosophy is ecosophy or ecological wisdom. He goes on to point out that ecophilosophy's purpose is to explore a diversity of human-nature interrelationships so as to foster "deeper and more harmonious relationships between place, self, community and the natural world" (Drengson). Arne Naess defined ecosophy as follows:

> … a philosophy of ecological harmony or equilibrium. A philosophy as a kind of sofia (or) wisdom, is openly normative, it contains both norms, rules, postulates, value priority announcements and hypotheses concerning the state of affairs in our universe. Wisdom is policy wisdom, prescription, not only scientific description and prediction. The details of an ecosophy will show many variations due to significant differences concerning not only the 'facts' of pollution, resources, population, etc. but also value priorities (Naess 1994, p. 124).

For many African communities, it would be wrong to over-consume resources and leave future generations with fewer means of survival. Thus for these societies, the current generation has moral obligations towards future generations as they are morally considerable (Naess 1994, p. 124; Arnold 2011, pp. 1–15).

According to indigenous African thought, the present generation is under obligation to the past (the living dead) and future generations. In Africa, people have direct and indirect obligations to future generations. In the first place, the land belongs communally to past, present and future generations. Individuals may privately possess livestock and movable possessions (Van Rooy 1997). The current generation ought to be grateful to, and follow the example of, past generations with regard to leaving behind a healthy land. This indicates an indirect moral obligation towards future generations. According to Kevin Gary Behrens, "It is likely that this backward-looking notion of duties to posterity is one of the most significant contributions African thought can make to our conception of moral obligations to the future" (Behrens 2012, p. 187).

In the spirit of the common good, it is worth noting that African ethics does not require the current generation to pay attention to the rights and interests of future individuals, but rather those of *future communities*. Thus, the present generation can have obligations towards whoever will constitute members of communities in future generations. Consequently, those countries that irresponsibly aggravate climate change ought to learn from the ethical principle which highlights the intergenerational dimensions of global climate change and pay attention to the well-being of future generations over and above their current interests (Kelbessa 2014).

African environmental ethics recognises the proper place of non-human animals in the world, and the relatedness among human and non-human beings on the one hand and the environment on the other. Like human beings, non-human animals have the right to exist. Africans are committed to the preservation of all species (Kelbessa 2014).

The African common good ethic has very well found expression in the *Ubuntu* philosophy. The philosophy of ubuntu can also offer useful lessons for human beings. The concept of ubuntu, found in the Bantu languages of East, Central and Southern Africa, captures the essence of what it means to be human (Munyaka and Motlhabi 2009; Murove 2004; Ramose 2002; Tutu 1999). It affirms that a person is a person through other persons (*I am because we* are)—the conception of persons is centrally constituted by and through their relations to others. These relations inform a person's understanding of moral behaviour and how to live in a community (Munyaka and Motlhabi 2009, p. 82).

Furthermore, and very critical is the fact that, the principle of ubuntu can be extended to the biotic community. Ubuntu transcends individualism which is common in the Western world. Its ethical message is not restricted to one's own ethnic group, regional community or nation, but can cover all of humanity and the more-than human world. This standpoint is succinctly stated by Ramose. He stresses that the fulfilment of natural duty includes caring for one another and for the natural environment (Ramose 2002, p. 124; Chibvongodze 2016, pp. 157–166).

4 African Eco Relationalism

In light of the discussion in the foregoing section, African thoughts on the environment are usually described as *holist*. Oruka, and Juma speak of '… a holistic outlook in which everything is related to everything else. This interrelatedness requires a corresponding philosophical approach that looks at nature in its totality…' (Oruka and Juma 1997). Opoku writes '…in the African perspective, the community and its members as well as its environment constitute a complex whole…' (Opoku 1993). Peterson describes the relationship between humans and nature as 'complex, holistic and dialectical' (Peterson 2004). Kelbessa goes so far as describe the worldview of the Oromo people, the subject of his study, thus: 'To some degree, Oromo ethics is close to Leopold's land ethic…' (Kelbessa 2005). Given the fundamental communitarianism of African ethics, it is hardly surprising to discover similarities with Western holist perspectives.

After all, an ethic that values harmonious relationships between humans and others, particularly, living aspects of nature, based in the belief that everything in nature is interdependent and interrelated, is bound to share some characteristics of holist environmental ethics. Both reject anthropocentricism, and both emphasise a fundamental interdependence in nature, and they also often seem to share a common language, and employ similar analogies (Behrens 2012, p. 67).

One example of how Western holists use similar analogies to those used in African thought is Aldo Leopold's famous claim that right actions are those tending to protect 'the integrity, stability and beauty of the biotic community' (Leopold 1949). The use of the phrase 'biotic community' is clearly reminiscent of African conceptions of nature as a community or family. Callicott points out that much holistic thought regarding the environment is predicated on the modern scientific understanding of nature and ecology, and uses similar language in doing so: 'Now the general world view of the modern life sciences represents all forms of life on planet Earth both as *kin* and as fellow members of a social unit – *the biotic community*' (Callicott 1986). African worldviews seem, then, to employ similar analogies to those used by holists. When nature is understood as a community or family, value is attributed to the community itself *as well as* to the individuals making up a community. Furthermore, it is pertinent to note that Western holistic conceptions are strongly grounded in the findings of ecological science, that organisms and species exist in ecosystems and are mutually interdependent, where damage to one part of the system can severely affect other parts of the system, and so forth. In a pre-scientific way, African indigenous thought recognises a similar interrelatedness in nature, albeit in more metaphysical terms, such as sharing a life force.

Given these similarities in conception, language and analogy, it might seem *prima facie* appropriate to describe this trend in African environmentalism as holist. Despite the similarities, I now show that the African emphasis on relationality is, in fact, distinct from holism, and that it is the distinguishing characteristics that offer the most promise for ecological integrity. In making this distinction, it is imperative to articulate an African-inspired environmental ethic, a relational, harmony-prizing ethic, which shall be referred to in this chapter as 'African Eco-Relationalism'.

One of the most striking things about much African thought regarding moral obligation is its refusal to firmly prioritise either the interests of individuals or the interests of communities. Eze writes: 'The relationship between the individual and community is dialogical for the identity of the individual and the community and is dependent on this constitutive formation. The individual is not prior to the community and neither is the community prior to the individual' (Eze 2008). The individual is fully realised only within the community. In terms of personhood, true humanity is achieved only through harmonious relationships with others, which, as I have explained, are characterised mainly by solidarity, friendship, care and seeking the common good. Since the interdependence of everything in nature is acknowledged, this community is understood to extend beyond human beings, and to encompass all natural entities.

Humans cannot be fully realised without recognition of their belonging to the community of nature, and behaving accordingly, demonstrating respect for nature through promoting harmonious relationships in this natural community. Interdependence is a fact, and harmonious relationships with not merely persons, but a variety of natural beings, as well, are to be sought as the way to self-realisation.

Despite his seemingly low estimation of traditional African values as a source for a promising environmental ethic, Callicott did recognise the potential of this strand of African thought:

> …far more vividly than in the modern Western worldview, individuality is not only counterbalanced by community identity but one's unique individuality is defined in part by one's social relationships and expressed through social interaction…. In this notion of embedded individuality – of individuality as a nexus of communal relationships – we may have the germ of an African environmental ethic. Add to the intense sense of social embeddedness an equally vivid sense of embeddedness in the biotic community, and anthropocentric African communitarianism might then be transformed into a non-anthropocentric African environmentalism (Callicott 1986, pp. 166–167).

Kevin Behrens rightly argues that Callicott somehow missed a '…vivid sense of embeddedness in the biotic community' in at least some African thought. He argues that African communitarianism can embrace all of nature, and thus provide a promising African environmentalism, based on respect for nature, individual objects in nature, and the nurturing of harmonious relationships and solidarity with other natural things (Behrens 2012, p. 70).

It is the sense of 'embedded individualism' within communitarianism that sets African Relational Environmentalism apart from Western expressions of holism. Leopold's and Callicott's 'Land Ethic' places final value in the community of nature, or the 'land', understood broadly as the biospheric community. One of the persistent challenges to holism is that if one accords ecosystems and nature final value, respect for humans becomes secondary to the needs of the biosphere. Could this not imply a moral imperative, for instance, to cull human beings in the interests of the survival of the earth? Such conclusions are very counter-intuitive. Callicott has made various attempts to counter this objection, claiming that morality itself has developed in an evolutionary way, gradually extending the boundaries of the 'community' of moral considerability from family to clan, to tribe, to nation, to all persons, and ultimately to nature (Behrens 2012, p. 70).

As long as holists give primacy to the interests of the biosphere, it is hard to see how they can overcome the problem that their view might be said to be misanthropic. Where African eco relationalism differs is that it does not prioritise the biotic community over members of the community, or vice versa. Right actions are not those that tend to promote the integrity and stability of the land, they are those that promote harmonious relationships between us and other members of the community of nature. One cannot promote harmonious relationships and completely ignore the interests of the individuals involved in the relationships. Wiredu writes: 'To adjust the interests of the individual to those of the community is not to subjugate one to the other. The relationship is purely … symmetrical. We could just as easily have described it as the adjustment of the interests of the community to those of the individual' (Wiredu 2008). While nature may sometimes need to take precedence over the interests of some individuals, this does not imply an unconditional primacy of the biotic community over individual interests. Sometimes humans may be morally obliged to give priority to, say, the preservation of an endangered species, perhaps by not using part of its habitat for agriculture, but this does not imply that human interests must always defer to those of the biosphere. An African sense of relationality respects both the interests of individuals and those of the community of nature, giving ultimate primacy to neither, a controversial claim that deserves to be taken seriously.

5 Implementing African Eco Relationalism for Ecological Integrity

Protecting the interests of one element of the natural community to the detriment of others will be counterproductive to achieving ecological integrity. Contemporary writers such as Jennifer Nedelsky (Nedelsky 1993, 1998, 2008) and Joseph William Singer (Singer 2000) have progressed the relational view of rights and interests. Nedelsky, for example,[1] argues that rights should not be viewed as clashing of individual interests or as absolute power within predefined spheres. Instead she situates rights within a broad web of relationships, limited by their impact on others She notes, 'what rights in fact do and have always done is construct relationships—of power, of responsibility, of trust and obligation' (Nedelsky 1993, p. 13). From this perspective, rights create a setting in which individuals and communities live their lives and interact with others. This setting consists partly of rules requiring individuals to respect the legitimate interests of others. Other rules are designed to ensure that the comprehensive Earth community functions well.

Rights and interests should thus be understood as socially constructed, involving not only relations between people, but also between people and things. While some might view relationships with nature as a limitation on human autonomy, Nedelsky points out that individuals achieve autonomy not in isolation, but by a combination of independence and dependence. She argues that this approach shifts the focus from protection against others to structuring relationships so that they foster autonomy; this makes some of the most basic presuppositions about autonomy to change: dependence will no longer be seen as the antithesis of autonomy but a precondition in the relationships. Consequently, interdependence becomes the central fact of political life (Nedelsky 1993, p. 8).

Placing interests of members of the natural community in a relational context has an important consequence for environmental preservation because it contextualises

[1] Although Nedelsky's argument refers to human relationships, the term 'relationship' will be extended in this thesis to include nature, having established above that non-human entities can be right holders.

and places limits on human property rights. It will be implicit in such a framework that property owners will look to nature as the standard or measure for their action. Consider, for example, that the Nigerian Government passed a law which recognised that the water bodies in Nigeria had the right to a healthy flow of water. Plainly such a law would conflict with the pre-existing rights of landowners, farmers and towns to draw water to meet their own needs. On a purely individualistic account of rights, this dispute would focus on the adversarial clashing of interests and a legal hierarchy would be established to settle the dispute. If the focus is on relationship, then both social and environmental contexts become relevant to resolving the dispute. A court could examine their respective needs and their specific relationship. From a conservation perspective, a good outcome would be that the irrigators' right to draw water would become encumbered. This does not remove the irrigators' right; it contextualises its use and requires careful consideration and knowledge of the needs of the river and the unique function it plays in the ecosystem. Water could be drawn *to the extent that these vital functions are maintained.*

Setting realistic, measurable and locally relevant biodiversity objectives will provide a sound basis for the negotiation of trade—offs that would result in the implementation of eco relationalism. It is important to recognize that local people may have their own priorities for Nature that differ from those of outside conservation groups. Building on these may provide a sound basis for securing local buy—in (Sheil et al. 2006). It is also important to recognize that sustainable use of biodiversity may be a more attractive option for local people than total protection. If people can benefit from using a species they are more likely to conserve it. This goes to question the reality of either pure anthropocentrism or ecocentrism. Aldo Leopold realized this paradox and wrote: 'conservation is a state of harmony between men and land' (Leopold 1970). Ideally, both sides must look towards a compromise. Grumbine argues that if ecosystem management is to take hold and flourish, the relationship between the new goal of preserving ecological integrity and the old standard of providing goods and services for humans must be reconciled (Grumbine 1994).' Jason Veil also made the same argument by asserting that a balanced approach to ecosystem management is a call to face reality, gather input from opposing viewpoints, and let these lessons drive future policy objectives. This depends on utilitarians recognizing the folly of relentless consumption and environmentalists accepting the reality of society's continued growth (Veil 2001). What is advocated therefore is 'sustainable development in an environmentally and socially-responsible manner through discussion, redesign, and active negotiation with potentially-affected communities' (Caron 2003).

6 Conclusion

Unlike mainstream Western ethics, African environmental ethics has recognized the interconnectedness and interdependence of all beings and the more-than human world.

African ecophilosophy on the impetus of African eco relationalism therefore holds promise for ecological integrity on the continent and beyond. This strand of African thought differs from biocentric positions in one very important respect. Biocentrism is usually characterized as an individualist position, locating moral value in particular properties of individual organisms or in maximising utility for such individuals. The African eco relationalism is strongly grounded in a communitarian value system. It prizes fostering harmony in the community of all living things, stressing their interrelatedness and bondedness.

To be an object of moral concern, rationality, intelligence, and language are not required, although different beings have different mental capacities and roles. The unity of the whole establishes an ethical obligation for human beings toward nature. Africa has different cultures that have helped to shape positive moral attitudes toward the natural environment and its human and nonhuman components. Africa eco-philosophy therefore has the potential to promote ecological integrity on the continent and beyond. An African emphasis on the interrelatedness or interconnectedness of everything in nature is the foundation for a morality that prizes harmonious relationships as against moral consideration criteria proposed by holists, ecocentrists, sentientists, and biocentrist individualists. In fact, it is able to expand this sphere even further essentially communitarian and relational, and thus able to embrace the most useful aspects of both.

Emphatically, the most promising aspect of this approach is its refusal to prioritize individuals over the community or family of nature, or vice versa.

References

Arnold D (2011) Introduction: climate change and ethics. In: Arnold D (ed) The ethics of global climate change. Cambridge University Press, pp 1–15

Avila N, Carvallo J, Shaw B, Kammen D (2017) The energy challenge in Sub-Saharan Africa: a guide for advocates and policy makers. Oxfam Res Backgrounder:25–26

Awolalu Ọ (1976) Sin and its removal in African traditional religion. J Am Acad Relig 44(2):275–287

Behrens K (2012) Moral obligations towards future generations in African thought. J Global Ethics 8(2–3):179–191

Callicott B (1986) The search for an environmental ethic. In: Regan T (ed) Matters of life and death: new introductory essays in moral philosophy. Random, New York, p 265

Caron C (2003) Defining the public interest, negotiating rights: the influence of environmental impact on assessment legislation in Sri Lanka. Paper submitted for the 9th International Conference on Sri Lanka Studies, Matara, Sri Lanka, 28–30 November 2003

Chibvongodze D (2016) Ubuntu is not only about the human! An analysis of the role of African philosophy and ethics in environment management. J Hum Ecol 53(2):157–166

Drengson A, Ecophilosophy. Ecosophy and the deep ecology movement: an overview. http://www.ecopherics.net/pages/DrengEcophil.html. Accessed 14 Feb 2019

Ekwunife A (1990) Consecration in Ibo traditional religion. Jet Publishers, Enugu, p 30

Eze M (2008) What is African communitarianism? S Afr J Philos 27:386–399

Fariga A, Ipimoroti M, Chirenje T (2018) Environmental pollution in Africa. Environ Dev Sustain 20(1):41–73

Grumbine E (1994) What is ecosystem management? Conserv Biol 8(1):31
IPBES (2018) Africa: loss of biodiversity puts current and future generations at risk. Assessment Report on Biodiversity and Ecosystems Services for Africa
Kelbessa W (2005) The rehabilitation of indigenous environmental ethics in Africa. Diogenes 52:17–34
Kelbessa W (2014) Can African environmental ethics contribute to environmental policy in Africa? Environ Ethics 36(1):31–61
Leopold A (1949) A Sand County Almanac. Oxford University Press, Oxford, pp 224–225
Leopold A (1970) A Sand County Almanac. Sierra Club/Ballantine, p 295
Munyaka M, Motlhabi M (2009) Ubuntu and its socio-moral significance. In: Murove F (ed) African ethics: an anthology of comparative and applied ethics. University of KwaZulu-Natal Press, Scottsville, pp 63–84
Murove F (2004) An African commitment to ecological conservation: the Shona concepts of Ukama and Ubuntu. XLV Mankind Q:195–215
Naess A (1994) Deep ecology. In: Merchant C (ed) Key concepts in critical theory: ecology. Humanities Press, New Jersey, p 124
Nedelsky J (1993) Reconceiving rights as relationship. Rev Const Stud 1:1
Nedelsky J (1998) Reconceiving autonomy: sources, thoughts, and possibilities. Yale J Law Fem:86
Nedelsky J (2008) Reconceiving rights and constitutionalism. JHR 7(2):139
Omofonmwan SI, Osa-Edoh GI (2017) The challenges of environmental problems in Nigeria. J Hum Ecol 23(1):53–57
Opoku K (1993) African traditional religion: an enduring heritage. In: Olupona J, Nyang S (eds) Religious plurality in Africa. Mouton de Gruyter, Berlin
Oruka O, Juma C (1997) Ecophilosophy and parental earth ethics. In: Oruka O (ed) Philosophy, humanity and ecology. ACTS Press, Nairobi, p 117 quoted from Kevin Behrens, "An African Relational Environmentalism and Moral Considerability" (2014) Perspectives in African Environmental Ethics and Philosophy. 36(1), 65
Penny R, Desertification and deforestation in Africa. Encyclopedia of life support systems <https://www.eolss.net/Sample-Chapters/C19/E1-05-06-09.pdf>
Peterson R (2004) Central African voices on the human-environment relationship. In: Gottlieb R (ed) This sacred earth. Routledge, New York, p 174
Ramose M (2002) African philosophy through Ubuntu, revised edition. Mond Books, Harare
Sheil D, Puri R, Wan M, Basuki I, Heist V, Liswanti M, Rukmiyati N, Rachmatika I, Samsoedin I (2006) Recognizing local people's priorities for tropical forest biodiversity. Ambio 35(1):17
Singer J (2000) Entitlement: the paradoxes of property. Yale University Press
Tutu D (1999) No future without forgiveness. Rider and Random House, London
Van Rooy JA (1997) Scriptural ethical principles and traditional African ethics. In die Skriflig 31(1&2):93–106
Veil J (2001) Anthropocentrism and ecocentrism: a balanced approach (April 2001). www.courses.psu.edu/for/for410_mem14/Essays/Spr_01/Jason_Veil.PDF. Accessed 16 Oct 2019
Vidal J (2016) Air pollution more deadly in Africa than malnutrition or dirty water, study warns. The Guardian, October 20, 2016
Wiredu K (2008) Social philosophy in postcolonial Africa. S Afr J Philos 27:332–339

State Global Responsibility for Environmental Crises: The Ethical and Legal Implications of a State's Failure to Protect Human Rights

Kathryn Anne Gwiazdon

Abstract This chapter seeks to explore the interconnectedness of life, the threats to that interconnectedness, and the responsibility of those in power for harms against humanity. It will do this through an ethical and legal lens, understanding that in both ethics and justice, engagement and discourse is key to evolution and enlightenment, democracy and diplomacy. This chapter opens with an exploration of ethical engagement, and how it can be used as a practical tool to address the greatest harms we face. It will use human rights as a focal point for its analysis of state global responsibility, for if our human constructs (or governance systems) do not protect humanity, there is a clear and pervasive need for fundamental change. Ultimately, we must ask ourselves, is the current structure of our state and global governance sustainable, is it protecting and nurturing humanity and human civilization? As a way forward, the chapter proposes extending the ethical and legal concept of ubuntu from human relationships to state relationships. It will argue that the principles and values behind ubuntu should be extended to the state and that the true, relational aspect of state sovereignty must be brought to the forefront instead of the harmful hyper-individualism fostered by hyper-nationalism.

1 Introduction

"A human being is a part of the whole called by us universe, a part limited in time and space. He experiences himself, his thoughts and feeling as something separated from the rest, a kind of optical delusion of his consciousness. This delusion is a kind of prison for us, restricting us to our personal desires and to affection for a few persons nearest to us. Our task must be to free ourselves from this prison by widening our circle of compassion to embrace all living creatures and the whole of nature in its beauty." (Sullivan 1972).

K. A. Gwiazdon (✉)
Center for Environmental Ethics and Law, Tampa, FL, USA
e-mail: kathryngwiazdon@environmentalethicsandlaw.org

L. Westra et al. (eds.), *Ecological Integrity in Science and Law*,
https://doi.org/10.1007/978-3-030-46259-8_10

During his tenure at Princeton University, Albert Einstein wrote these words of comfort to a friend who was grieving the loss of his daughter. Although a simple note to a friend, it gives us much insight into Einstein's life philosophy: the interconnectedness of life; the artificial and partisan limitations we place on ourselves in the community of life; and a way out of the darkness—out of the self-inflicted harm—by expanding our circles of care.

Einstein was a well-known internationalist and pacifist and was ardently opposed to racism (or human partisanship), nationalism (or state partisanship), and militarism (or state protection or state harm) (Sullivan 1972; Ghodsee 2015). It was the rise of Hitler, however, and his policies and philosophies of *kriegmenschen* (war people) and *schwertglauben* (sword believers), that he came to believe that pacifism was not absolute, and that those seeking peace cannot close their eyes to reality (Ghodsee 2015). He understood that racism and nationalism can destroy life on earth and can only be prevented if world leaders (or those with power) set aside conflicts, and "consider yourselves only as members of a biological species which has had a remarkable history, and whose disappearance none of us can desire" (The Russell-Einstein Manifesto 1955). In one of his final acts, and alongside 10 other scientists calling for the end of nuclear proliferation and war, he wrote, "The abolition of war will demand distasteful limitations of national sovereignty… Remember your humanity, and forget the rest." (The Russell-Einstein Manifesto 1955).

This chapter will explore the interconnectedness of life (humans and nature, humans within nature, and humanity as part of nature), the threats to that interconnectedness (i.e. nationalism), and the responsibility of those in power (here, nation-states) for harms against humanity. It will do this through an ethical and legal lens, understanding that in both ethics and justice, engagement and discourse is key to evolution and enlightenment, democracy and diplomacy, as opposed to absolute maxims forced upon our diverse cultures and communities of life. After all, you do not need to know perfect justice to know injustice; and you do not need to know what perfect ethics is, to advance ethical action (Sen 2009).

This chapter will open with an exploration of ethical engagement, and how it can be used as a practical tool to address the greatest harms we face. What methodology can be used to highlight ethical principles in action? What process can be used to highlight the underlying values that guide all decisions? It will use human rights as a focal point for its analysis of state global responsibility, for if our human constructs (or governance systems)—at their most basic level—do not protect humanity, there is a clear and pervasive need for fundamental change.

This chapter will not provide a detailed overview of law and policy concerning human rights and the environment, as this has been done extensively elsewhere, with numerous commitments over numerous decades. What is important to note, however, that despite these "commitments," grave human rights violations continue to occur and are increasing in frequency. We are in the midst of global environmental crises that threaten all humanity: mass-scale biodiversity loss, the acidification and choking pollution of our oceans, and climate change all lead to human insecurity, food and economic insecurity, and national and global insecurity (Gwiazdon 2018). Why is this happening, despite our calls to halt these harms?

To begin to understand, we must go deeper than data; we must look into humanity and how we—or those in power—structure ourselves. Ultimately, we must ask ourselves, is the current structure of our state and global governance sustainable? Is it protecting and nurturing humanity and human civilization?

To adequately address our global environmental crises, we must address our global governance crises, and to do that, we must look at the power-holders in global governance: the nation-state (state). States have the power to create, implement, and enforce laws that guide the behavior of their citizens, industries, and relationships with other nations, and it is this power that creates their coinciding responsibility to not knowingly create or allow harm with those laws and relationships. Yet in our current governance system, it appears that states have the power, or even the right, *to* harm. In a world of growing hyper-nationalism and the dominant statist role of the state, grave harm is being done to all humanity in defense of some humanity. We need to redefine how states relate to one another.

As a way forward, the chapter proposes extending the ethical and legal concept of *ubuntu* from human relationships to state relationships. *Ubuntu* is an ethic—or a set of values—of care and interdependence. It roughly translates into, "I am because we are" and places our identity, our humanity, within our relationships to others (Tutu 1999). This is not at the expense of the diversity of local people, places, and cultures, but through that diversity. The principles and values behind *ubuntu* should be extended to the state; after all, a state is only a state because of other states. The true, relational aspect of state sovereignty must be brought to the forefront instead of the harmful hyper-individualism fostered by hyper-nationalism.

2 A Methodology for Ethical Engagement

Ethics is the foundation for law and governance, and all decisions have ethical underpinnings. Ethical engagement looks to the underlying values of our decisions. What do we value, or not value—and why? What are our priorities as citizens and as a society—and why? And how does corruption—of people and institutions—manipulate ethical action? Ethical engagement is more than creating a list of shared principles that guide life in a community of life, although that is not unhelpful, but rather it is a questioning of the values that lead to decisions that guide every aspect of our lives.

Ethical engagement can be as diverse as humanity, and so it naturally fosters openness and creativity in methodology, discourse, and decision-making. For example, in human rights discourse, human rights are considered "strong ethical pronouncements as to what *should* be done" (Sen 2009). They are not already established legal rights, but rather an "invitation to initiate some fresh legislation" that asserts the "critical importance of certain freedoms" and the corresponding need to "accept some social obligations to promote or safeguard these freedoms" (Sen 2009).

The U.S. Declaration of Independence was seen as the moral standard for which the country should strive to attain. When Thomas Jefferson drafted the document in

1776, he proposed a philosophy of human rights inherent in all people, "We hold these truths to be self-evident, that all men are created equal, that they are endowed by their Creator with certain unalienable Rights, that among these are Life, Liberty and the pursuit of Happiness…" (U.S. Declaration of Independence 1776). And then U.S. President Abraham Lincoln made the document—this moral standard—into the centerpiece of his policies, as witnessed most notably in the Gettysburg Address of 1863, when he argued that the Declaration of Independence is a statement of principles through which the U.S. Constitution should be interpreted (U.S. President Lincoln 1863). It was ethical engagement that highlighted the principles of the Enlightenment, and it was ethical engagement that then translated those ethical principles into law and governance.

Ethical engagement helps us to highlight harms, the perpetrators and the victims. Philosophy has a crucial role in "bringing more discipline and greater reach to reflections on values and priorities, as well as on the denials, subjugations, and humiliations from which human beings suffer across the world" (Sen 2009). John Rawls saw "justice as fairness," and that the principles of liberty and equality should be applied to protect the least advantaged among us (Rawls 1951; Sen 2009). He pointed to the "moral powers that people have, related to their capacity for a sense of justice and for a conception of the good" (Sen 2009).

Ethical engagement allows us to address the root causes of our crises, and to analyze the values that led to our crises. For example, Don Brown believes that instrumental rationality is the greatest challenge to ethical thinking and acting (Brown 2018). Instrumental rationality is operationalized throughout law and governance; it looks at what is most efficient or cost-beneficial, absent of the values involved. Immanuel Kant (1724–1804) saw the harm of instrumental rationality, and made the connection between morality and governance. He believed that political moralists "fashion their morality to suit their own advantage as a statesman" (Brown 2018). They "wield power on Earth…" by "advocating might instead of right" and "[making their] principles subordinate to [their] end" (Kant 2009).

Jürgen Habermas saw "the colonization of the lifeworld" occurring due to our instrumental systems, which he defined as money and power, and that, "[The] tendency of the system to colonize the lifeworld leads to greater fragility and to disequilibrium or instability… and crises in the system" (Finlayson 2005). "As the domain of the lifeworld shrinks… social pathologies arise:

1. Decrease in shared meanings and mutual understanding (anomie—or a lack of ethical standards in an individual or group)
2. Erosion of social bonds (disintegration)
3. Increase in people's feelings of helplessness to take responsibility for their actions and for social phenomenon (demoralization); and
4. Destabilization and breakdown in social order (social instability)" (Finlayson 2005).

Lifeworld, a concept first used by German philosopher Edmund Husserl (1859–1938), is the everyday world we share with others. It is the "informal and unmarketized domains of social life: family and household, culture, political life outside of organized parties, mass media, voluntary organizations, etc." (Finlayson 2005). The lifeworld is gradually and constantly changing" and "is the medium of the symbolic and cultural reproduction of society". Our systems of money and power are predatory on our lifeworld and "cut deep channels into the surface of social life" where "agents fall naturally into pre-established patterns of instrumental behavior" (Finlayson 2005). They are "guided by their role into patterns of action in pursuit of financial aims" (Finlayson 2005). And Habermas warned that "systems of money and power steer agents towards ends that are not related to understanding or consensus" (Finlayson 2005).

In "Outline of a Decision Procedure for Ethics," Rawls explored whether a procedure was available to adjudicate and prioritize competing interests (Rawls 1951). He believed that a judge should be "intelligent, well informed, have adequate time to make his decision, be sane, as opposed to suffering from mental illness, and not have personal interests in the case" (Rawls 1951). And Amartya Sen believes that open impartiality and critical scrutiny are central features of "the general framework of ethical and political evaluation" (Sen 2009). Unfortunately, these do not provide much guidance on a clear and consistent methodology for ethical engagement, something that could be put into practice in law and policy. But perhaps the reason why a mechanistic approach is largely absent for ethical engagement is because a mechanistic approach is not applicable, or appropriate.

3 Is State Sovereignty Sustainable?

Human rights are concerned with "the whole of humanity (*universorum*), united in Earthly society and distributed in national groups" (Tutu 1999). Therefore, violations of human rights are violations to human harmony, to society-building, to stability, security, and governance—to life, and the foundations of life, itself. They harm present and future opportunity, present and future generations, as well as other species, individuals, and states. They also target and prioritize harm to those who often need the most protection: vulnerable peoples, such as women, children, minorities, indigenous peoples, and refugees, and in the case of the environment, vulnerable species and habitats. Harm to humanity, harm to ourselves and our communal whole—which includes harming the foundations of life—gives insight into who we are. It shows what we are capable of, and through the reasons we give in defense of harm, our priorities, our values, and where our circles of care *end*.

3.1 The Particular Power of the State in Our Societal Relationships

States are the power-holders in structuring society, at all levels, local to global. The decision-makers within a state create the structures and laws that govern all actors within a society, within and among nations: what air we breathe, what water we drink, what food we eat, who can be imprisoned, who can have their life taken away. Through the rule of law, government leaders determine and regulate good and bad behavior, what is considered harm and how much is allowable. Therefore, it is the values of those in power that determine our overall well-being.

The particular—and arguably singular—power of the state is perhaps most evident by the lack of an adjudication and enforcement body at the global level for state harm. A governance structure that creates, implements, and enforces state global responsibility is simply not in place, and so the crises continue, placing human integrity, environmental integrity, societal integrity, ecological integrity in jeopardy.

In the Fifth Assessment Report of the Intergovernmental Panel on Climate Change (IPCC), Chapter 4 argued for the need for equity considerations in climate change mitigation, yet found that "The fact that there is no universally accepted global authority to enforce participation is taken by some to mean that sovereignty, not equity is the prevailing principle," and that this approach has "perverse results" (Fleurbaey et al. 2014). The Chapter 4 authors do believe that "governments can choose to act on moral rather than purely self-interested principles" (Fleurbaey et al. 2014) But will they?

In most states, the powers of the state derive from their citizens. Philosophically, this is done through social contract theory; legally this is done through governance structures (Gwiazdon 2018). When states fail to protect human rights, including protecting the environment, this is a dereliction of the state's duty to protect its citizens, and the duty to protect is the most important duty of the state in its social contract with each citizen (Gwiazdon 2018). The corresponding duty not to harm, however, is one that extends beyond citizenship, to all humanity.

Within the U.N. framework, states do have a Responsibility to Protect (R2P) that extends outside their borders. This commitment was endorsed by all member states at the 2005 World Summit. However, it is narrowed to only include genocide, war crimes, ethnic cleansing, and crimes against humanity.[1] The original drafters envisioned a wider scope of application, which included, "overwhelming natural or environmental catastrophes, where the state concerned is either unwilling or unable to cope, or call for assistance, and significant loss of life is occurring or threatened" (International Commission on Intervention and State Sovereignty 2001). This was limited in scope by heads of state and governments at the 2005 World Summit limited its scope, arguing that to include "climate change or the response to natural disasters, would… stretch the concept beyond recognition or operational utility" (The Secretary General. Report on the Secretary General on Implementing the Responsibility to Protect 2009).

What is promising is that the principle is based in human rights and global responsibility, i.e. that with sovereignty comes responsibility to protect all populations from certain crimes. Sovereignty is not an allowance to harm, and it

[1] See generally http://www.un.org/en/genocideprevention/about-responsibility-to-protect.html [Accessed 2018-11-10].

does not abdicate one from responsibility. The ability to act under R2P is also limited, however, as it is enforceable only by the U.N. Security Council, whose inability to respond is noted below.

3.2 Human Rights: State Versus Humanity

Albert Schweitzer believed "Ethics is nothing other than reverence for life. Reverence for life affords me my fundamental principle of morality, namely, that good consists in maintaining, assisting and enhancing life, and to destroy, to harm or to hinder life is evil" (Schweitzer 1923). His views are largely based on his experiences while doing medical missionary work in colonial Africa, where he witnessed some of the worst crimes against humanity in human history,

> Oh, this 'noble' culture of ours! It speaks so piously of human dignity and human rights and then disregards this dignity and these rights of countless millions and treads them underfoot, only because they live overseas or because their skins are of different color or because they cannot help themselves. This culture does not know how hollow and miserable and full of glib talk it is, how common it looks to those who follow it across the seas and see what it has done there, and this culture has no right to speak of personal dignity and human rights...

And, "I will not enumerate all the crimes that have been committed under the pretext of justice" (Schweitzer 1993).

Humboldt also saw the links between humanity, morality, and extended that into our treatment of the natural world, "the natural world was linked to the 'political and moral history of humanity,' from imperial ambitions that exploited colonial crops to the migration of plants along the paths of ancient civilizations" (Wolfe 2015). He "looked at the plants in the context of the wider world and with a focus on revealing global patterns, he connected colonialism, slavery, and economics" and saw that, "colonialism was disastrous for people and the environment" because "it was based on inequality" (Wolfe 2015). He understood that "everything was interrelated: climate, soils, and agriculture with slavery, demographics, and economics. It was European barbarity that had created this unjust world" (Wolfe 2015).

From colonialism to war, and colonialism as war, from racism to nationalism, many calls for ethical action have been the result of witnessing great harms against humanity. The U.N. was created in 1945, after the atrocities of World War II, its charter "reaffirmed faith in fundamental human rights, and dignity and worth of the human person" and committed all member states to promote "universal respect for, and observance of, human rights and fundamental freedoms for all without distinction as to race, sex, language, or religion" (Charter of the United Nations 1945).

Three years later, the U.N. detailed those human rights in the Universal Declaration on Human Rights (UDHR), and the ICCPR and the ICESCR were seen as a legally-binding, unpacking of those principles. In addition to these founding human rights documents, other key agreements have highlighted the protection of vulnerable peoples from the cruelty of mankind, covering discrimination, women, children, migrant workers, and indigenous peoples. Most environmental treaties, agreements, and programs also include human rights considerations, such as public health, development, and equity, such as the CBD, UNFCCC, and such programs as the UN SDGs, Agenda 21, and Harmony with Nature.

The agreements, programs, and people of the human rights movement are vast—yet are they working? Richard Falk, an international law expert and former U.N. Special Rapporteur on the Situation of Human Rights in the Palestinian Territories Occupied since 1967 (2008–2014), believes that all of these developments "suggest the emergence of human rights, but not yet its global acceptance as enforceable law" (Polychroniou 2018). Decades of documents and diplomacy, and we are merely at "the emergence of human rights".

The procedural history of the UDHR even shows that "it would not have received widespread support from leading governments had it been negotiated as a lawmaking treaty with the effect of eroding sovereign rights" (Polychroniou 2018). In today's U.N., Secretary General António Guterres believes that a power struggle among the Security Council, the U.N.'s permanent, veto-holding body, has prevented them from addressing many of our current global crises, "'the Security Council doesn't correspond anymore' to today's international power dynamics… Those powers have struggled—and in many cases failed—to take meaningful collective action on almost all the pressing issues of the day: the suspected genocide against Rohingya Muslims in Myanmar, Russia's seizure of Crimea, the disaster in Syria" (Friedman 2018).

All laws—and the punishments for breaking them—are in place to prevent or deter harm to others. Yet where are the punishments for state's violating their ethical and legal duties not to harm others? Enforcement largely belongs to the courts, and in human rights law, and outside of national implementation, there are several bodies, yet limited by procedural or subject matter jurisdiction, or even ability to enforce its decisions.

One of the most powerful bodies capable of enforcing human rights law is the International Criminal Court (ICC). Yet its mandate is narrow, major states are not parties to its Rome Statute, and they have been criticized for bias. In 2016, the Office of the Prosecutor of the ICC even expanded the mandate to include environmental crimes, but there has yet to be any action particular to environmental harms (Sarliève 2018).

Looking at these courts, it seems that even when state actions cause harms, there is no reliable and consistent global justice system that can hold all states accountable for harms. Therefore, it seems largely up to citizens and civil society organizations, who are often powerless, with the monumental tasks of demanding protections for human rights from their and other states, placing environmental harms within human rights, demanding accountability when those rights are violated, and demanding a change in their government if the harm continues. Falk believes that, "It is civil society that has tried to keep the 'human' side of human rights as integral to the protective mission, while governments are again limiting their view of rights to the 'rights' side as an entitlement of a national citizenry, especially those who are native born" (Polychroniou 2018). And so, the harms often continue unabated.

3.3 A Denial of Reality, an Embrace of Injustice

Great harms often bring calls for great peace—for the unity of mankind. When President Lincoln delivered the Gettysburg Address, he highlighted America's national purpose that all people are created equal, and that "government of the people, by the people, for the people, shall not perish from the earth" (U.S. President Lincoln 1863). We are in a period of great harm now, and before we can look into how we call for great peace, we must understand the root causes of this harm. Are our current state and global governance bodies "of the people, by the people, for the people"; and if it is not, is this why we are perishing from the Earth?

Kant believed that a pacific federation (*foedus pacificum*) among states could seek to end all wars, as opposed to a peace treaty that may end one war ("which are actually only truces"), and from there an international state (*cevitas gentium*) would "continue to grow until it embraced all the peoples of the Earth" (Kant 2009). However, he saw that it was not yet the will of the states to adopt a world republic, or a "cosmopolitan constitution," and so, "they reject *in hypothesi* what is true *in thesi*" (Kant 2009).

"The issue of borders is as much a moral question as it is one of policy calculation" and states have effectively set arbitrary lines to defend acts of aggression or allow for inaction (Smith 2017). Borders may exist to define nations and protect citizens from outside harms, but borders are also used to deny the extraterritoriality of harms or to deny refugees. Why does an artificially-made disconnect with our neighbors make harm tolerable or justifiable? We must "look for alternatives for the sake of the West's own moral and legal integrity" (Smith 2017).

Authoritarianism is "the enforcement or advocacy of strict obedience to authority at the expense of personal freedom; a lack of concern for the wishes or opinions of others".[2] After the U.S. pulled out of the Climate Change Accord, and before they pulled out of the Human Rights Council, they were attacking the very norms of relational, global governance. Former U.S. Ambassador to the U.N., Nikki Haley stated "...there are multiple things that we have gotten out of... that were beneficial to other countries but not to us: the Paris Climate agreement, we got out. The Iran deal, we've gotten out of. UNESCO was another one that we got out of. And so you are seeing the efforts of defunding those things that are not helpful to us and that... aren't in the United States interest. And the Human Rights Council and the Office of the Human Rights Commissioner we are going to make targeted cuts. If there are certain things that are not beneficial to our interests..., we're going to get out of it" (Speech: Nikki Haley 2018).

Is statist sovereignty to global governance as authoritarianism is to national governance? In global governance, with no enforceable justice system, are states saying that harm to the "other," the non-citizen, is simply allowed? We see this in the

[2] Authoritarianism, defined. Oxford Dictionaries. https://en.oxforddictionaries.com/definition/authoritarian [Accessed 2018-11-10].

killing of journalists, with no international response; we see this in the destruction of the Amazon and indigenous peoples, with no international response.

And authoritarian, nationalist regimes are on the rise among states. Is this due to weaknesses in global governance or state governance? U.S. President Donald Trump is a self-described nationalist who promotes a "sovereignty-first" agenda (Baker 2018). He "avoids bringing up uncomfortable human-rights issues with foreign leaders in the belief that he can get a better deal by leaving thorny subjects and the internal affairs of other nations off the table" (Kounalakis 2018).

Falk argues that,

> …nationalist goals are always given precedence without a second thought. From these perspectives international law should not be allowed to interfere with fidelity to a nationalist agenda. At the same time, the pressures exerted by migratory flows stemming from war torn regions, especially the Middle East and Africa, and from ecologically challenged habitats, have weakened mainstream support for human rights, and especially for those who are refugees or asylum seekers. Despite neoliberal globalization, and in some respects, as a reaction to it, the state system has become more statist, with a corresponding retreat in efforts to protect the human rights of vulnerable peoples, especially if they are regarded as strangers to the community of the nation state. (Polychroniou 2018)

Under this hyper-nationalism, a double-harm, or exacerbation of harms, also occurs. States can cause or be contributors to global crises, like political instability, poverty, mass migrations, or climate change, and then shield themselves with state sovereignty and hyper-nationalism. They cause global crises, yet then exacerbate that harm by continuing that harm, closing their borders, severely limiting or refusing refugees, and/or cutting humanitarian aid. The victims of human and environmental harms then also become the victims of the statist state.

4 Ubuntu: A Relational Ethic for a Relational World

"A person is a person through other persons." (Tutu 1999)

And a nation is a nation through other nations. From Africa came the origins of our lives as humans, and from Africa could come the path to our future, the migration of ubuntu (Posth et al. 2017). Ubuntu is an ethical and legal principle, rooted in southern African, that recognizes that we are humans through other humans, due to our communal and harmonious relationships with one another. Ubuntu is an ethic of interdependence, "We belong in a bundle of life" and "my humanity is caught up, is inextricably bound up, in yours" (Tutu 1999).

And ubuntu unites by acknowledging our differences. According to Michael Onyebuchi Eze, "A person is a person through other people" strikes an affirmation of one's humanity through recognition of an 'other' in his or her uniqueness and difference," and "induces an ideal of shared human subjectivity that promotes a community's good through an unconditional recognition and appreciation of individual uniqueness and difference" (Eze 2008, 2010). Therefore, actualizing an ethic of ubuntu does not lose individual identity or culture, but rather, is conditional upon those differences. Tutu believes, "We are different so that we can know our need of one another, for no one is ultimately self-sufficient" (Metz 2017).

The ethic of ubuntu can be extended beyond humanity, as well, into our connections and interdependence with all life. Motlatsi Khosi, a lecturer in African philosophy and ubuntu at the University of South Africa, argues that "[*ubuntu*] also relates to the way people interact with both the natural and metaphysical worlds, the latter consisting of unseen elements such as ancestors and God... It should be about how to make sense of the connection with the universe, animals, nature and humans" (Chibba 2013). Khosi also believes that living ubuntu is "a reaction to the dehumanising world of individualism, materialism and isolation" (Chibba 2013). It is a tool for reconciliation from grave harms.

Let ubuntu be the ethic of interdependence that combats extreme individualism and extreme competition—the roots of inequality. Let ubuntu be the ethic of covenant and community against those seeking excessive power. Let ubuntu be the ethic of care against corruption; the ethic of harmony against disharmony.

We may learn much in ethics, law, and global governance from South Africa's apartheid and their transition into a democracy. "Apartheid not only prevented 'races' from identifying with one another or exhibiting solidarity with one another. It went further by having one 'race' subordinate..." and allowing that race to harm others (Metz 2017). It made people "less human for their failure to participate on an evenhanded basis and to share power, wealth, land, opportunities, and even themselves" (Metz 2017). This provides uncomfortable parallels with how states globally govern today. Some have more power than others and use that power to cause harm or shield themselves from responsibility. And some are more vulnerable, and often made-so at the hands of the powerful, such as with colonialism and climate change. Opportunities are unequal, power is unequal, life is unequal. Is our current system of global governance a model of global apartheid?

5 Conclusions

Our current model for state sovereignty and global governance is not sustainable. We are bearing witness to this fact. We need a new methodology that can adequately address the global crises we face today. The values, ethics, and structures that harm humanity can no longer be embraced.

Let us extend the ethic of ubuntu to the community of states; let us replace statist sovereignty with relational sovereignty. A human is a human because of, and among, other humans; humanity is defined by, empowered by, constrained by, and conditional upon others, and our relationships to others. A nation is a nation because of, and among, other nations; sovereignty is defined by, empowered by, constrained by, and conditional upon others. And in no man-made social construct can one harm another without consequence.

If our governance structures—created by our underlying ethics—are not livable, if our decisions are causing global crises that threaten all life, then we must seek new ways forward. Let not a construct of humanity—the state—destroy humanity.

References

Baker P (2018) 'Use That Word!': Trump embraces the 'nationalist' label. New York Times. 23 October 2018. https://www.nytimes.com/2018/10/23/us/politics/nationalist-president-trump.html. Accessed 01 Dec 2018

Brown D (2018) Interview with Gwiazdon K. via Zoom. 28 November 2018

Charter of the United Nations (1945). http://www.un.org/en/charter-united-nations/. Accessed 2018-11-10

Chibba S (2013) Ubuntu is about relationships, 19 September 2013. https://www.brandsouthafrica.com/people-culture/people/ubuntu-is-about-relationships. Accessed 2018-11-10

Eze MO (2008) What is African communitarianism? Against consensus as a regulative ideal. S Afr J Philos 27(4):386–399

Eze MO (2010) Intellectual history in contemporary South Africa. Palgrave MacMillan

Finlayson G (2005) Habermas: a very short introduction. Oxford University Press

Fleurbaey M et al (2014) Sustainable development and equity. In: Edenhofer O et al (eds) Climate Change 2014: Mitigation of Climate Change. Contribution of Working Group III to the Fifth Assessment Report of the Intergovernmental Panel on Climate Change. Cambridge University Press, Cambridge

Friedman U (2018) Is democracy dying? U.N. Secretary-General: American Power Is in Decline, the World Is 'in Pieces'. 13 September 2018

Ghodsee KR (2015) Einstein's pacifism: a conversation with Wolfram Wette. Institute for Advanced Study. https://www.ias.edu/ideas/2015/ghodsee-einstein-pacifism. Accessed 2018-12-01

Gwiazdon K (2018) The state versus the environment: the ethical and legal implications for state non-action in protecting the foundations of life. In: Westra L, Bosselmann K, Gray J, Gwiazdon K (eds) Ecological integrity, law and governance. Routledge

International Commission on Intervention and State Sovereignty (2001) The Responsibility to Protect: Report of the International Commission on Intervention and State Sovereignty. December 2001. http://www.globalr2p.org/media/files/iciss_report.pdf. Accessed 2018-11-10

Kant I (2009) An answer to the question: what is enlightenment? (trans: Nisbet HB). Penguin Books, London

Kounalakis M (2018) Under Trump's "sovereignty doctrine," foreign tyrants have nothing to worry about. Miami Herald. 01 November 2018. https://www.miamiherald.com/opinion/op-ed/article220974935.html. Accessed 2018-11-10

Metz T (2017) What Archbishop Tutu's ubuntu credo teaches the world about justice and harmony. The Conversation. 4 October 2017. http://theconversation.com/what-archbishop-tutus-ubuntu-credo-teaches-the-world-about-justice-and-harmony-84730. Accessed 2018-11-10

Polychroniou CJ (2018) Human rights, state sovereignty, and international law: an interview with Richard Falk. Global Policy Journal. 11 September 2018. https://www.globalpolicyjournal.com/blog/11/09/2018/human-rights-state-sovereignty-and-international-law-interview-richard-falk. Accessed 2018-11-10

Posth C et al (2017) Deeply divergent archaic mitochondrial genome provides lower time boundary for African gene flow into Neanderthals. Nat Commun 8, Article number: 16046. https://www.nature.com/articles/ncomms16046. Accessed 2018-11-10

Rawls J (1951) Outline of a decision for ethics. Philos Rev 60(2)

Sarliève M (2018) Can criminal courts help save the environment? 10 October 2018. https://www.justiceinfo.net/en/justiceinfo-comment-and-debate/opinion/39189-can-criminal-courts-help-save-the-environment.html. Accessed 2018-11-10
Schweitzer A (1923) Civilization and ethics. A. & C. Black, London
Schweitzer A (1993) Reverence for life, 2nd edn. Irvington Publishers, New York
Sen A (2009) The idea of justice. The Belknap Press of Harvard University Press, Cambridge
Smith N (2017) A world without borders: Richer, Fairer, and more free. Foreign Affairs. 28 February 2017. https://www.foreignaffairs.com/articles/world/2017-02-28/world-without-borders. Accessed 2018-11-10
Speech: Nikki Haley (2018) Foundation for defense of democracies: presentation of the Jeane J. Kirkpatrick Award to U.N. Ambassador Nikki Haley. 28 August 2018. https://s3.us-east-2.amazonaws.com/defenddemocracy/uploads/documents/FDD-Summit-Ambassador-Haley.pdf. Accessed 2018-11-10
Sullivan W (1972) The Einstein Papers. A man of many parts. New York Times. 19 March 1972, https://www.nytimes.com/1972/03/29/archives/the-einstein-papers-a-man-of-many-parts-the-einstein-papers-man-of.html. Accessed 2018-11-10
The Russell-Einstein Manifesto. 9 July 1955. Available at http://umich.edu/~pugwash/Manifesto.html. Accessed 2018-12-01
The Secretary General. Report on the Secretary General on Implementing the Responsibility to Protect. U.N. Doc A/63/677. 12 January 2009. https://documents-dds-ny.un.org/doc/U.N.DOC/GEN/N09/206/10/PDF/N0920610.pdf?OpenElement. Accessed 2018-11-10
Tutu D (1999) No future without forgiveness. An image book. Doubleday, New York
U.S. Declaration of Independence (1776). https://www.archives.gov/founding-docs/declaration-transcript. Accessed 2018-11-10
U.S. President Lincoln, A. Gettysburg Address. 1863. http://rmc.library.cornell.edu/gettysburg/good_cause/transcript.htm
Wolfe A (2015) The invention of nature: Alexander von Humboldt's new world. Vintage Books, New York

Promoting Human Attachment to Place in Ecological Law

Geoffrey Garver

Abstract Contemporary law, including environmental law, is built on worldviews and moral philosophies that encourage and sanction detachment of people from place. As ecological law matures, it must embody principles for re-embedding people in the places that nourish them and for constraining or abolishing remote decisions and policies made without intimate connection to, concern about and knowledge of the local people and places affected. Time-tested Indigenous worldviews and ways of life, as well as principles of eco-cultural restoration, provide insights on how to promote attachment to place in the law.

1 Introduction

In her brilliant book *Braiding Sweetgrass,* Robin Wall Kimmerer tells of asking a friend what place "you know best and knows you in return," to which he replied, "My car" (Kimmerer 2013, p. 125). Years later, she writes, his car is where he tried to kill himself. You detach yourself from the Earthly places that nurture you at your peril. *Braiding Sweetgrass* is a compelling reflection on the beauty and wonder of attachment to place, how to maintain and cherish it and, ultimately, why it is an essential element of an enduring human presence on this finite planet. Yet, Kimmerer's profound and resonant message has not yet come through in any significant way in the domain of law. Indeed, legal regimes around property, state sovereignty, trade, land use, energy, agriculture, finance—it is hard to find an area of law that does not fit—predominantly favor detachment of people from place and decisions that are taken without attachment to or knowledge about the remote people and places they affect, far too often negatively.

G. Garver (✉)
Concordia University, Department of Geography, Planning and Environment, Montreal, QC, Canada

McGill University, Economics for the Anthropocene/Leadership for the Ecozoic Partnership, Law and Governance Research Initiative, Montreal, QC, Canada

L. Westra et al. (eds.), *Ecological Integrity in Science and Law*,
https://doi.org/10.1007/978-3-030-46259-8_11

As the still developing field of ecological law matures, it will be important to ensure that it embodies principles that promote re-embedding people in the places that nourish them and that severely constrain or abolish remote decisions and policies made without intimate connection to, concern about and knowledge of the local people and places affected. Contemporary law, including environmental law, too rarely gets this balance right. In this chapter, the many ways in which contemporary law is detached from place in harmful ways will be reviewed and the potential for ecological law to reverse course will be explained.

2 Detachment from Place in Contemporary Law

Contemporary law, including environmental law, is built on worldviews and moral philosophies that encourage and sanction detachment from place. Building on themes of human superiority over all other beings in classical Greek and Judeo-Christian thought, the ideological separation of humans and nature, with humans superior to and accorded mastery over nature, firmly took hold of normative constructs like economics and law in the Enlightenment and now dominates thought systems governing the human-nature relationship around the world (Capra and Mattei 2015). The exceptions—indigenous worldviews across the globe, Eastern thought, the theology of St. Francis of Assisi, for example—are important potential antidotes and seeds for transformation, but they have had little effect reining in the key drivers of catastrophic ecological change in the modern globalized economy.

The historical record of humanity's social metabolism, in particular the growing divide between where people live and the often remote ecosystems that sustain them, aligns with this human-nature ideological binary. Especially in Europe, the early modern era from around 1400 to 1800 was a period of externalization of the means of providing for a growing population as local sources became fully or over-exploited (Richards 2003). European colonization of much of the Americas, Africa and Asia in this period was a matter of finding both new sources of material and energy—including slaves—to sustain Europeans at home and new places for Europeans to settle. The spirit of conquest and domination with which European monarchs undertook to colonize faraway places, sometimes relying on papal bulls they read as granting permission to assert sovereignty over "discovered" non-Christian lands with a view to converting any inhabitants to Christianity (*Romanus Pontifex* 1455, *Inter Caetera* 1493), largely precluded any real notion of attachment to the places and people there. The quasi-legal privateers who took part in this phase of land and resource grabbing evolved into today's legal persons called corporations, many of whom continue to engage in modern, globalized land and resource grabbing (Korten 2009). Condemnation of the brutality of colonization began as early as the sixteenth century, notably by the Spanish Dominicans Bartolomé de las Casas in Mexico and Francisco de Vitoria in Spain (Behr 2010), and the Doctrine of Discovery is increasingly disavowed or repudiated now (*Tsilhqot'in Nation v. British Columbia* 2014). Nonetheless, the detachment of people from place and nature that expanded considerably in the early modern era remains deeply entrenched in societies across the globe.

Most likely, the majority of people in colonized places are largely unaware of how they still benefit from, and their legal systems still remain connected to, the legacy of the brutal colonization of already inhabited lands during the early modern era. In the United States, Columbus Day remains an official holiday to celebrate the "discovery" of the inhabited Americas by one of history's most renowned and brutal land and resources grabbers (Zinn 1980), despite its occasional renaming as Indigenous Peoples' Day. Apologies for colonization and its offshoots, like residential schools where First Peoples in Canada and the United States were compelled to renounce their cultural heritage (Kimmerer 2013), are rarely if ever matched with action anywhere near sufficient to right the wrongs that colonization exacted on countless people and places.

Legally authorized neo-colonial activities that continue to carry forward the historical record of detachment from place are widespread across the globe. *The Land Grabbers* (Pearce 2012) lays out a daunting survey of contemporary land and resources grabbing, including grabs for long-term leases for export-oriented and biofuel agriculture in many African countries, palm oil plantations in Indonesia, soy and cattle farms in Brazil and more. His account includes "green grabs" in Africa and Patagonia in which lands with centuries old histories of successful integration of human communities in their supporting ecosystems are taken over for human-free parks and wilderness areas—a notion of conservation that fully aligns with human-nature dualism. One of the first large-scale green grabs was the establishment of Yellowstone National Park, which required the removal of twenty-six Native American tribes who used the land for various purposes (National Park Service 2019).

A future with colonies on Mars, extractivist missions as in the movie *Avatar* and other fantasies reminiscent of the virtually treeless cartoon series *The Jetsons* is the logical terminus of the still prevailing mindset grounded in detachment from place. Ultimately, this mindset, rooted in Descartes, Bacon and others' refinement of core Judeo-Christian ideologies in the Enlightenment, is about "'mastering' nature to reinvent Eden" (Merchant 2003). The legal foundations of this techno-optimist, Cartesian future are already being laid. For example, in 2015, the United States enacted the *U.S. Commercial Space Launch Competitiveness Act*, which gives American citizens property rights to any "asteroid resource or space resource" they obtain. Luxembourg enacted a similar law in 2016, just before investing heavily in an asteroid-mining company (Consensys Space 2019). Although the *Outer Space Treaty of 1967* states that "outer space is not subject to national appropriation by claim of sovereignty, by means of use or occupation, or by any other means," the United States and Luxembourg claim the treaty does not preclude giving private entities rights to space resources (Hamer 2017). Even if asteroid mining has not yet happened, missions by the United States, the former Soviet Union, Russia, the European Union, India, Japan and China to explore Mars have been ongoing since 1960 (Hamer 2017). NASA has a well-funded Mars Exploration Program and notes that the red planet “could someday be a destination for survival of humankind” (NASA 2019).

Space colonies as imagined in *the Jetsons*, centered on a nuclear family with mother, father, their children and a dog, essentially consist of an expansion of the kind of false attachment to place associated with cookie-cutter suburban, car-dependent communities made up mostly of single family homes. Attachment to these kinds of places, which the American dream and the social status associated with home ownership encapsulate, is hard to reconcile with the kind of attachment to place and the communities of diverse and flourishing life that are associated with a human-inclusive ecocentric worldview and ecological law. The superficial attachment to place of the American dream ultimately perpetuates a profound detachment from a truly ecological notion of place.

Remote policymaking is a close cousin of remote ownership. Indeed, remote ownership and land and resource grabbing can only exist upon a legal and policy foundation that allows for them. In many legal systems, policymakers and public officials must assess environmental, social and cultural impacts before approving policies, programs or decisions that affect faraway people and places. However, laws requiring analysis and consideration of such impacts are almost invariably procedural only, leaving the policy and decision makers free to select options that do not avoid or minimize impacts (Lazarus 2012). Further, cumulative, indirect, vague and uncertain impacts are often excluded or only marginally incorporated in impacts assessments, such that a complete and holistic picture of the full range of impacts is rarely if ever achieved, even when assessments cover broad strategies or policies (Garver 2013). Typically, moreover, political and economic pressure to proceed with policies or decisions with impacts on remote people and places, despite whatever impact assessments reveal, is considerable—or else the decision or policy would not even be under consideration. In short, impact assessment has hardly reigned in the remote harm policy and decision making often causes.

Geoengineering to address the climate crisis is an extreme extension of the ideology of detachment from place, whether by ownership or policy. Geoengineering typically involves grand technological projects to completely master and control the Earth’s climate system, so that all corners of the planet would be subject to centralized decisions made inevitably without anything close to full knowledge of impacts on local people and ecosystems (Minteer 2012). Moreover, many geo-engineering proposals require continual management in order to maintain their potential to mitigate global heating (Robock 2008). Obviously, contemporary impact assessment approaches are incapable of ensuring adequate assessment of the impacts on local communities and ecosystems globally of geo-engineering proposals such as placing orbiting mirrors in space to deflect sunlight or seeding the oceans with iron filings to promote carbon storage in algae (Robock 2008).

Table 1 summarizes ideological underpinnings of anthropocentric contemporary law that support ongoing detachment of people from place, and how human-inclusive ecocentric ecological law offers counterpoints by which attachment or re-attachment to place might be realized.

Table 1 Values of contemporary law and ecological law

Contemporary law: detachment from place	Ecological law: attachment to place
Anthropocentrism	Ecocentrism
Colonialism and remote ownership	Indigeneity and localism
Growth and development	Resilience and regeneration
Efficiency	Sufficiency
Monetary (single-criteria) valuation	Multi-criteria valuation
Commodification	Respect & reciprocity
Instrumental view of nature	Relational view of the life community
Rational individuals	Diverse and pluralistic communities
Monocultures	Diversity and ecological integrity
Selfishness	Sharing and solidarity

3 Attachment to Place in Ecological Law

Kimmerer writes that "[f]or the sake of the peoples and the land, the urgent work of the Second Man may be to set aside the ways of the colonist and become indigenous to place" (Kimmerer 2013, p. 207). But, she explains, inviting settler societies to become indigenous to place does not mean letting them "take what little is left" (Kimmerer 2013, p. 211). It is instead an invitation to "give your gifts and meet your responsibilities [and] to live as if your children's future matters, to take care of the land as if our lives and the lives of all our relatives depend on it" (Kimmerer 2013, p. 215). Indigenization and attachment to place according to Kimmerer are a way to fend off "Windigo thinking"—the worldview of a mythical Anishinaabe monster driven entirely by self-interest and an insatiable appetite for excess. Kimmerer associates Windigo thinking with humanity's current dominant commitment to "progress" and development founded on infinite economic growth, privatization, commodification and invariably a detachment from place. *Braiding Sweetgrass* teaches that attachment to place, founded on respect for all life and deep appreciation of a reciprocal relationship with the Earth and its life community, is key to a more hopeful vision of humans embedded within the community of life of which they are members. "Gratitude is a powerful antidote to Windigo psychosis" (Kimmerer 2013, p. 377).

This notion of attachment to place must be central to any conception of ecological law. Yet, the previous section reveals that Windigo psychosis is deeply entrenched in human society across the globe. Shifting to ecological law that fosters attachment and reattachment to place in a way that avoids perpetuating the power dynamics of domination and colonialism is a monumental task and perhaps nearly unimaginable. Ultimately, this shift will entail a radical rethinking of centuries old worldviews and the institutions and societies founded on them that see humans as separate from and superior to all other life. Nonetheless, the vision of ecological law is ultimately positive and life affirming, and seeds of its emergence can already be found. Indeed, contrary to the papal bulls noted above, Pope Francis's encyclical *Laudato Si'* provides such an opening by calling into question modern anthropocentrism and advocating for ecological citizenship and ecological conversion of both individuals and communities (*Francis* 2015).

Giving ecological limits paramount importance and treating humans as part of nature are key features of ecological law (Garver 2013), but ecological law also must include elements of justice related to protection of local communities of life, including the people embedded in them. Ensuring interhuman, interspecies and intergenerational fairness is a baseline in ecological law from which further notions of ecological justice derive (Garver 2013; Sbert 2019). Yet, ecological law also embraces the kind of flexibility and diversity associated with ecological resilience (Gunderson and Holling 2002), and therefore a plurality of ways by which to ensure justice within and across bioregional communities of life is most appropriate. Locally crafted systems of fairness and justice can be found in the examples of sustainable governance of common pool resources that Ostrom and others have described (Ostrom 1990; Agrawal 2001). The strength of these models is that the systems of fairness and justice involved emerged from within the communities and have withstood the test of time.

The 2007 United Nations Declaration on the Rights of Indigenous Peoples (UNDRIP) contains elements of the kind of ecological justice that must be part of ecological law. For example, it guarantees indigenous peoples rights to self-determination and self-governance. As well, it requires free, prior and informed consent of indigenous peoples before they may be removed from their lands, before laws affecting them are adopted or implemented, and before projects affecting their lands are approved. These requirements go beyond the duties of governments to consult meaningfully with indigenous peoples with regard to decisions affecting them, but not to obtain their consent, as in Canada (*Chippewas of the Thames First Nation v. Enbridge Pipelines Inc.* 2017). However, ecological law would go further than UNDRIP in requiring justice to extend the logic of UNDRIP more clearly to future generations and to other species, for example by prohibiting all human activities that make possible catastrophic global heating that imperils future generations of all life.

Attachment to place should also be enhanced and protected in ecological law through incorporation of principles of relational justice. Under ecological law, the "othering" and instrumentalization of nature and non-human life reflected in contemporary law, for example by framing nature primarily as a provider of ecosystem services to humans, would give way to recognition of the intrinsic value of all elements of ecosystems and notions of intersubjectivity between all members of the Earth's life communities (Michelot and Aseeva 2017). Any notion of ecosystem services to humans would have to be paired with an equivalent notion of human services to ecosystems (Garver 2019), akin to the concept of ecological solidarity integrating

"(1) territorial solidarity; (2) social solidarity; [and] (3) interconnexion between humans and nature" (Michelot and Aseeva 2017).

Although ecological law must include these justice principles to reflect mutuality and reciprocity within Earth's life communities, it must also allow for human life to flourish within those communities. Ecological law cannot preclude all human activity that affects nature. The key is to shift from the emphasis in contemporary economics and law on satisfying desires and generating profits through efficiency and growth to a rediscovery of the indigenous wisdom of sufficiency and taking no more than what is truly needed (Kimmerer 2013). Enough, not more, is the guiding principle (Capra and Mattei 2015). To achieve this, ecological law should contain the spark of competition and markets within an engine of cooperation, solidarity and community (Brown and Garver 2009). Ecological economics and the degrowth movement provide many ingredients and recipes for doing so: for example, progressive taxation, cooperative enterprises instead of profit-driven corporations, direct democracy, guaranteed basic income, valorizing care work, producing and consuming locally, promoting sharing, and increasing the commons (Kallis 2018).

4 Attaching or Re-attaching to Place in an Altered World

Humanity has no choice but to take the world as it exists now and to work from there toward a desired future. Taking the world as it is means accepting that places and societies exist that humans have altered over the expanse of history. Cities and other dense human settlements that have sprouted around the globe since the Neolithic transition 10,000 years ago are an obvious example. Altered places also include novel ecosystems altered irreversibly by human activities, yet now with structures and functioning that support life in new but ongoing ecological relationships (Seasted et al. 2008). People long ago turned vast forests in Scotland, Iceland, Newfoundland and other northern places into tree-scarce heaths, bogs and peatlands, some with an astonishing level of biodiversity that did not exist in the time of forests (Huntley et al. 1997; Hambrecht 2015). Through human agency, exotic species have firmly established themselves across the globe, sometimes with devastating effects on the ecosystems they invaded, and often there is no going back. Starlings in North America, stoats in New Zealand, brown snakes in Guam, and rabbits in Australia are some infamous examples.

Looking to the future, a concept like ecological integrity must be broad enough to include at least some of these altered places and societies, especially where they are, or with ecological or eco-cultural restoration viably could be, life-enhancing and vital (Garver 2019). The essential element of any meaningful and enduring attachment or re-attachment to place is a deep commitment to relating to the community of all life of the place as a life-affirming, life-enhancing and respectful member. This appreciation and awareness of what gives a place its unique and sacred identify is at the heart of the sense of communion on which this notion of membership is based. Allowing this posture of awe and reverence only where nature is pristine and relatively free of humans is no longer possible, and so it is an attitude to nurture in even the most altered corners of Earth.

Eco-cultural restoration is an approach to make the best of Earth's mosaic of altered and relatively unaltered places. As an applied science, eco-cultural restoration necessarily involves intentional interventions with a goal in mind (Garver

2019). To be consistent with ecological law, eco-cultural restoration applied at landscape and bioregional scales should consist of a weaving together of historical fidelity, ecological integrity and community engagement in a quest to invigorate place-based ecological solidarity with reference to intimate local knowledge of place and a forward-looking sense of what seemed to get the balance right in the past (Garver 2019).

5 Conclusion

That some human communities and individuals still retain deep respect for and attachment to the life systems that sustain them gives reason to be hopeful about the human prospect on the finite planet Earth. Yet, attaching or re-attaching humanity to the ecosystems and communities of life of which people are a part on a broader scale will be an enormous challenge. The ideologies of separation of humans from nature are widespread and deeply entrenched in much of the world, and in much of the law. Moreover, strong arguments exist that these ideologies are ultimately rooted in the emergence of agriculture, which invariably requires an intentional manipulation and control of ecosystems to serve human ends and allow permanent settlements. Farming has been described as the original human act of colonization (Fischer-Kowalski and Haberl 1997). It always limits biodiversity, and modern industrial agriculture does so extensively. Farming also led to the rise of civilizations and the earliest known forms of codified law (Prince 1904; Kornfeld 2009), as well as to the gradual replacement of the diversity of polytheist societies with the singularity of monotheist ones (Armstrong 1994). At the very least, parallels can be drawn between this religious transformation and the evolution toward the monoculture of value that money represents in modern society, the narrowing of the range of acceptable outcomes in engineered industrial systems where consistent product quality is essential for success, and modern notions of law as rigid and monolithic. They all tend to squelch life-enhancing diversity.

Expecting humanity to abandon agriculture and the entire harvest of socio-cultural norms and civilizational ordering that farming spawned would be futile and unnecessary. Full material and spiritual attachment or re-attachment to place in a way that gives hope for the human prospect will undoubtedly entail a new relationship with farming, one inspired, for example, by agroecology and the Three Sisters—corn, squash and beans—of the indigenous people of Turtle Island (Kimmerer 2013). The quest should be for restoration of the eco-cultural systems that predated or escaped the rise of the kind of agriculture associated with the rise of industrialization in a globalized, capitalist world. This transformation may well entail a reduction in the number of people integrated harmonistically into Earth's life communities. Malthus saw this as a harsh and dismal future (Sieferle 2001), but from the perspective of worldviews that celebrate strong spiritual connections between people and their inhabited places, it is a future of beauty, wonder and collective well-being for humanity and other life.

The implications for law of this spiritual transformation and reconnection are also wondrous. A strong secondary effect of the detachment of place and instrumentalization of nature and non-human life that permeates contemporary law is detachment of people from the law itself (Capra and Mattei 2015). In the modern globalized world, law is typically seen as developed, administered and imposed from afar by professional specialists. Ideally, ecological law, like the *sui generis* systems of place-based rules, customs and traditions of well-governed common pool resources (Ostrom 1990), will be an integral and cherished part of the communities and places that develop and apply it.

References

Agrawal A (2001) Common property institutions and sustainable governance of resources. World Dev 29(10):1649–1672

Armstrong K (1994) A history of god: the 4,000-year quest of Judaism, Christianity and Islam. Ballantine Books, New York

Behr H (2010) A history of international political theory. Palgrave MacMillan, New York

Brown PG, Garver G (2009) Right relationship: building a whole earth economy. Berrett-Koehler, San Francisco

Capra F, Mattei U (2015) The ecology of law: toward a legal system in tune with nature and community. Berrett Koehler, San Francisco

Chippewas of the Thames First Nation v. Enbridge Pipelines Inc. 2017. Supreme Court of Canada, 2017 SCC 41

Consensys Space. 2019. https://consensys.space

Fischer-Kowalski M, Haberl H (1997) Tons, joules, and money: modes of production and their sustainability problems. Soc Nat Resour 10(1):61–85

Francis (2015) Laudato Si'. United States Conference of Catholic Bishops, Washington, DC

Garver G (2013) The rule of ecological law: the legal complement to degrowth economics. Sustainability 5(1):316–337

Garver G (2019) A systems-based tool for transitioning to law for a mutually enhancing human-earth relationship. Ecol Econ 157:165–174

Gunderson LH, Holling CS (eds) (2002) Panarchy: understanding transformations in human and natural systems. Island Press, Washington DC

Hambrecht G (2015) The First European Colonization of the North Atlantic. Chapter 9. In: Leone MP, Knauf JE (eds) Historical archaeologies of capitalism, 2nd edn. Springer, New York

Hamer A (2017) Asteroid mining? It's not as crazy as it sounds. https://curiosity.com/topics/asteroid-mining-its-not-as-crazy-as-it-sounds-curiosity/

Huntley B, Daniell JRG, Allen JRM (1997) Scottish vegetation history: the Highlands. Bot J Scotl 49(2):163–175

Kallis G (2018) Degrowth. Agenda Publishing, Newcastle-upon-Tyne UK

Kimmerer RW (2013) Braiding Sweetgrass. Milkwood Editions, Minneapolis

Kornfeld IE (2009) Mesopotamia: a history of water and law. Chapter 3. In: Dellapenna J, Gupta J (eds) The evolution of law and politics of water. Springer, New York

Korten DC (2009) Agenda for a new economy. Berrett Koehler, San Francisco

Lazarus R (2012) The National Environmental Policy Act in the U.S. Supreme Court: a reappraisal and a peek behind the curtains. Georgetown Law J 100:1507–1586

Merchant C (2003) Reinventing Eden: the fate of nature in western culture. Routledge, New York

Michelot A, Aseeva A (2017) From ecosystem services to ecological solidarity. In: Westra L, Gray J, Gottwald FT (eds) The role of integrity in the governance of the commons. Springer, New York

Minteer BA (2012) Geoengineering and ecological ethics in the anthropocene. Bioscience 62(10):857–858

NASA (2019). http://mars.nasa.gov

National Park Service (2019) Historic Tribes, Yellowstone. https://www.nps.gov/yell/learn/historyculture/historic-tribes.htm

Ostrom E (1990) Governing the commons: the evolution of institutions for collective action. Cambridge University Press, New York

Pearce F (2012) The land grabbers: the new fight over who owns the earth. Beacon Press, Boston, MA

Prince JD (1904) The Code of Hammurabi. Am J Theol 8(3):601–609

Richards JF (2003) The unending frontier: an environmental history of the early modern world. University of California Press, Berkeley CA

Robock (2008) 20 reasons why geoengineering may be a bad idea. Bull At Sci 64(2):14–18

Sbert C (2019) El Salvador's mining ban and mining in Ontario's ring of fire from the lens of ecological law. Vermont Law Rev 43(3):517–548

Seasted TR, Hobbs RJ, Suding KN (2008) Management of novel ecosystems: are novel approaches required? Front Ecol Environ 6(10):547–553

Sieferle RP (2001) The Subterranean Forest: energy systems and the industrial revolution. The White Horse Press, Cambridge

Tsilhqot'in Nation v. British Columbia. 2014. Supreme Court of Canada, 2014 SCC 44

Zinn H (1980) A people's history of the United States. Harper & Row, New York

Protected Wilderness: Is It Still Possible?

Anne Venton

Abstract Wilderness is defined as a natural environment on Earth that has not been significantly modified by human activity. In the last half of the twentieth century, growth in world population and hyper globalization of industrial capitalism have resulted in a major degradation of the Earth's natural environment and deforestation. Large majorities of the public have shown little interest in arresting the degradation while political leaders have ignored it in order to promote economic growth. Indeed, the huge growth in the inequality of wealth in most democratic states over the last forty years is a measure of the decline in their democracies and the rise of de facto oligarchies serving the interests of wealthy minorities and a singular focus on the objective of economic growth. This is especially marked in the 20 nation states with common law systems that have no environmental provisions in their constitutions such as the right to a healthy environment. Nevertheless, there are two grounds today for some optimism for achieving wilderness protection. The first is the recognition of the extraordinary environmental rights revolution that has taken place since 1970 whereby 130 nation states have amended their constitutions with provisions to protect their environments thus taking environmental protection out of the hands of political leaders and giving it back to the public through the courts. Secondly, public awareness of climate change has greatly increased in the last ten years and public advocacy for addressing climate change is wide spread. Along with that is an expectation that political leaders must now seriously address climate change.

1 Introduction

What is Wilderness? It is a natural environment on Earth that has not been significantly modified by human activity (Wikipedia, Wilderness).

A. Venton (✉)
Ontario Institute for Studies in Education Alumni Association, University of Toronto, Toronto, ON, Canada

L. Westra et al. (eds.), *Ecological Integrity in Science and Law*,
https://doi.org/10.1007/978-3-030-46259-8_12

A concern for environmental protection has recurred in diverse forms, in different parts of the world, throughout history. For example, in the late nineteenth and early twentieth century, John Muir (1838–1914) was an influential Scottish-American naturalist, author, environmental philosopher, glaciologist, and early advocate for the preservation of wilderness in the United States. He helped preserve Yosemite Valley and his work impacted on the US Congress which passed legislation to create Yosemite National Park in 1890. Muir is described by his biographers as the Father of National Parks (Wikipedia John Muir).

"Following a devastating 1906 earthquake in San Francisco the city needed water and electricity. Even though different possibilities existed for supplying these, Gifford Pinchot, an American forester and politician, supported the economic groups who sought to build a dam in the Hetch Hetchy Valley in Yosemite National Park. John Muir, and all those who would protect the integrity of Yosemite Park, fought a media and political battle but in the end "ravaging commercialism" won the day and the dam was built. Pinchot's conservationism-oriented management of nature and his priority on human use of the planet's natural resources started down the road of the harmful damaging ideological utopia that today we call sustainable development (Bergandi 2017, pp. 180–181)."

Sustainable development today is a term that provides the illusion of a protected environment for future generations while rationalizing the destruction of the environment. Today, the prevailing destruction of the environment raises the question "Is it still possible to protect the wilderness?" I have modified the question slightly to "Is wilderness protection more an aspiration than a goal that can be achieved?" I will develop the argument that whether wilderness protection is an achievable goal or not depends on whether or not the goal can be taken out of the hands of the political representatives and enshrined as a constitutional right. The argument is based on the premise that a genuine democracy exists. I will look at the interplay between constitutional rights and values of political leaders in the context of changing economic conditions.

2 Grounds for Optimism in the Environmental Rights Revolution

In a 2018 *Globe and Mail* op ed article, international lawyer and Graham Fellow in the Faculty of Law at the University of Toronto, Maria Banda, reported on some positive trends. She stated that "An extraordinary legal revolution is unfolding around the world. Last month, in an historic ruling, Colombia's Supreme Court declared that the Amazon is a legal person with rights – to be protected, conserved and restored – and ordered the state to reduce deforestation. Further, she stated that this past year alone, from India to New Zealand, four rivers, two glaciers and a sacred mountain have been granted personhood. The Inter-American Court of Human Rights declared that the regional human-rights treaty protects the rights of the environment as such. U.S. municipalities are joining in (Banda 2018)."

2.1 David Richard Boyd's Book the Environmental Rights Revolution

Environmental lawyer David Richard Boyd's 2012 book *The Environmental Rights Revolution* traces three world wide developments since World War II that have led to the environmental rights movement around the world. The first was the United Nations' *Universal Declaration of Human Rights* in 1948 which included a variety of rights including rights to life/security, food, water, health, and an adequate standard of living. The second development was a warning of an impending environmental crisis presented in Rachel Carson's 1962 book, *The Silent Spring*. Subsequent analyses by scientists revealed the crisis of excessive climate change arising from man-made green house gas emissions. A specific human right related to this climate change crisis is the right to a healthy environment. The third development was a wave of new and amended constitutions in many of the world's 192 countries. "In fact, 130 constitutions in the world, including the overwhelming proportion of those amended or written since 1970, include a state obligation to protect the environment or the right to a safe, healthy and ecologically sustainable environment (Brown and Brown 2015, pp. 170–171)."

2.2 Environmental Rights Provisions in Constitutions

Today, societal actors in many countries are engaged in the process of attempting to transform today's unsustainable societies into more healthy and sustainable ones. This exercise includes the counterbalancing of property rights with the right to a healthy, sustainable environment for individuals. In this comparative study of constitutions in 192 countries world wide, Boyd notes that "Constitutional environmental provisions have substantially increased the public's role in environmental governance (Boyd 2012, p. 239)."

Boyd raises two specific questions about environmental rights provisions in constitutions:

- Do provisions of environmental rights in constitutions, and in particular, the right to live in a healthy environment, matter?
- Do they have observable legal and extra legal effects that contribute to improved environmental protection and reduce the risks to human health? Extra legal effects include the evolution of public opinion, attitudes, values and behaviour that may arise as a result of the existence of a constitutional rights provision.

Boyd argues that constitutions in democracies articulate values (or aims) that govern political and social discourse for generations to come. These values are protected by incorporating them as rights of individuals and groups. Courts are empowered through the process of judicial review to be the guardian of these rights so as to protect vulnerable groups against the tyranny of the majority (Boyd 2012, p. 5). This conception of constitutions is similar to that of Aristotle as explained in the following. The achievement of the objective of equality of justice requires that the democracy has a constitution and is governed by the rule of law. By the rule of law, Aristotle means a manner of governance in which the political leaders are constrained beforehand by principles [in the constitution] which are known to the citizen body. A political regime that defers to the rule of law is, in his opinion, superior to a regime in which the present generation alone is dominant (Tabachnick and Koivukoski 2011, p. 197, 26).

2.3 The Constitutional Right to a Healthy Environment

Practical experience with the Right to a Healthy Environment has had positive effects on environmental health. For example, constitutional rights have been used by public interest litigants to close gaps in environmental law including the courts in Nepal and Costa Rica that ordered governments to reduce air pollution and protect fisheries (Boyd 2012, p. 235). To take another example, constitutional provisions have promoted justice by providing minimal standards for the poor, ethnic and marginalized communities such as improved drinking water in Chacras de la Merced, Argentina (Boyd 2012, p. 238).

Another effect of the constitutional right to a healthy environment has been its influence on environmental legislation. Boyd notes that court decisions based on the right to a healthy environment represent a remarkable evolution for the judiciary whose focus for centuries has been the protection of private interests where the promotion of public interests was the exclusive domain of the legislature. Today, Court Orders have incorporated innovative mechanisms to ensure compliance, including third party supervision and mandatory progress reports to the court and substantial court-imposed fines for non-compliance (Boyd 2012, pp. 280, 281).

Of the twenty-three nations employing exclusively common-law systems, only three have environmental provisions in their constitutions (Boyd 2012, p. 51). They trail the rest of the world in constitutional recognition of social, economic and environmental rights. The twenty-three include the United States, the United Kingdom and Canada (with the exception of Quebec). As of today, there is no positive obligation at the national level on the Government of Canada to actually fulfill Section 7 (life, liberty and security of the person) of the Charter of Rights and Freedoms vis a vis environmental rights (Government of Canada 1982).

2.4 Tunisia's 2014 Constitution on Protecting the Climate for Future Generations

In January 2014, Tunisia embedded climate change into its constitution. Tunisia's constitution, praised as one of the most progressive in the region, now obliges the state to contribute to the protection of the climate for future generations. It also says that the state shall provide the necessary means to eliminate environmental pollution. By this action, Tunisia has not only given citizens the right to ask their government to deal with climate change, but also elevated the concept of climate change to one of international law. Moreover, it could open doors for one country to sue another where one country is causing harm to another country by not abating its emissions. Tunisia is the third country to include climate change in its constitution. Ecuador and Dominican Republic are others. Tunisia's parliament has also created a commission for sustainable development and for rights of future generations which reflects a concern about the impact of global warming.

A March 2013 World Bank Report said that climate change is already damaging people's livelihoods and well-being in Tunisia and, without action, it will deepen the already significant poverty and unemployment leading to food insecurity and political instability. As well, climate change is already impacting on people's livelihoods and well-being in Libya and Egypt and has played a role in the Arab Spring. Another implication of the changes found in Tunisia's constitution regarding climate change is that it could empower climate change refugees. Having climate change provisions in a constitution could be a relevant argument for those refugees seeking protection (Aulakh 2014).

If Canada's legislators introduced climate change provisions in Canada's constitution today, any of its current policies or legislation or violations that are inconsistent with the constitution could be taken to court and possibly struck down.

3 Pessimism About Constitutional Rights as Protections

Constitutional rights are not fool proof solutions for today's environmental problems. Two examples of failure are found in Brazil and the European Union.

3.1 Brazil's Bel Monte Dam Controversy

The Kapayo are an indigenous group in the Amazon area in Brazil. They have fought, and continue to fight, many battles with the climatic enemy. Brazil's twenty-five year old constitution includes sections which protect the lands and rights of the Kapayo territory—an area of rainforest covering twice the area of Nova Scotia—from loggers, miners, ranchers and dam builders. The Kapayo territory, the largest tropical reserve in the world, includes 9000 people in 34 villages spread across 11 million hectares. Although the Kapayo are considered indigenous role models for their successful defence of land and culture from modern encroachment, powerful new developments threaten to change life as they know it. A massive dam development on the Xingu River, a body of water that forms the centre of their existence and a proposal to change sections of Brazil's constitution is the climactic enemy (Brown 2014).

In 2005, the Brazilian Congress voted to revive the dam on the grounds that its energy was essential to the security of the rapidly growing nation. Claiming that the project's environmental impact statements were defective and that the region's indigenous people were not adequately consulted, Brazil's federal Public

Prosecutor's Office filed a series of lawsuits to stop the complex which essentially pits one branch of the government against the other. The cases went to the country's Supreme Court, but judgements have been deferred, and the construction of the Belo Monte Dam has been allowed to proceed.

In 2011, construction on the $14 billion Belo Monte Dam finally began. The complex of canals, reservoirs, dikes, and two dams is located on the northern part of the Xingu river, where the river makes a giant U-turn called the Volta Grande. The project, which will have a maximum generating capacity of 11, 233 megawatts and is slated to be completed in 2020, has divided the country between those who see it as a way to deliver needed energy and those who condemn it as a social, environmental, and financial disaster (Brown 2014).

3.2 The European Union

The European Union, once considered as leaders in addressing climate change, are now talking about diluting their clean-energy ambitions. A deep and lasting economic slowdown and persistently high prices for renewable energy are causing the European Union's policy making body to allow exploration for shale gas using the disputed technique known as hydraulic fracturing.

As well, the European Commission's energy and climate proposals will probably include binding targets of reducing emissions by 35 to 40% by 2030 but the targets will likely be applied across the EU rather than to individual nations. Some officials want to make the new targets for renewable energy non-binding. A spokesman for the German Renewable Energy Federation, which represents the wind and solar industry, called an absence of binding goals for renewables "totally disappointing" and said that it would contribute to a marked reduction in the pace of expansion of renewables. It appears that environmental concerns have slipped down the political agenda in the face of complaints about rising energy costs. As well, Germany's plan to close its nuclear power plants and build up renewable energy is running into problems. However, the German environmental minister still says, "Europe must remain a leader in climate protection (Reed et al. 2014)."

4 Growing Public Awareness of Climate Change in the 21st Century

As noted previously, one of the values of having constitutional guarantees regarding environmental rights are extra legal effects. These include the evolution of public opinion, attitudes, values and behaviour that may arise as a result of the existence of a constitutional provision. Since the publication of Boyd's book there has been an substantial growth in public awareness of climate.

4.1 Earth Day 2014

On Earth Day April 22, 2014, the U.S. Public Broadcasting Service (PBS) aired a documentary film across North America via satellite titled "A Fierce Green Fire: The Battle for a Living Planet". The film, which was directed by Oscar-nominated Mark Kitchell, had premiered at the Sundance Film Festival. In 2014, the film was aired as part of the PBS *American Masters* special. The film presents a history of the environmental movement and is structured into the following five acts. Each has a central story and protagonist.

- David Brower and the Sierra Club's battle to halt dams in the Grand Canyon
- Lois Gibbs and Love Canal residents' struggle against 20,000 tons of toxic chemicals
- Paul Watson and Greenpeace's campaign to save whales and baby harp seals
- Chico Mendes and the Brazilian rubber tappers' fight to save the Amazon rainforest
- Bill McKibben and the twenty-five year effort to address the impossible issue of climate change

Many media outlets and blogs picked up the film's story focusing on humankind's attitude towards the world's oceans. Paul Watson, who is an eco-crusader, whale warrior and one of the interviewees in Kitchell's film, said that the images of vast, floating piles of debris including plastic and discarded junk in the southern Indian Ocean have been etched into TV viewers' minds ever since the month before Earth Day when the search for missing passenger jet Malaysian Airlines Flight 370 began on March 8, 2014. Watson estimated that more than 100 million tons of plastic is floating in the Earth's seas. He added that films like Kitchell's opened people's eyes to what is going on. After that, it is up to viewers to decide whether to act. Their actions could range from something as simple as recycling and cutting back on personal waste to becoming an active campaigner for environmental protection. Watson is best known to casual viewers as the star of the Animal Planet Program, *Whale Wars* (Kitchell 2014).

4.2 David Suzuki's Documentaries

David Suzuki's documentaries on CBC television also reach out to millions across North America and raise awareness of climate change and the climate crisis and the need to protect Wilderness areas. "Beyond Climate" (2018) a documentary directed by Ian Mauro and narrated by David Suzuki shows how citizens living on Canada's west coast live with climate change that features burning forests, melting glaciers and acidic oceans (Mauro and Suzuki 2018).

4.3 *Anthropocene: The Human Epoch film*

The world premier of the documentary film, *Anthropocene: The Human Epoch,* was launched at the Toronto International Film Festival (TIFF) in September 2018 and subsequently screened at other film festivals including Sundance and Berlin. The film was produced by the award-winning team of Jennifer Baichwal, Nicholas de Pencier and Edward Burtynsky and took four years to make. It is the third in a trilogy that includes *Manufactured Landscapes* (2006) and *Watermark* (2013) that followed the research of an international body of scientists. After nearly 10 years of research, scientists are arguing that the Holocene Epoch gave way to the Anthropocene Epoch in the mid twentieth century because of profound and lasting human changes to the Earth. The film is a cinematic meditation on humanity's massive reengineering of the planet (Baichwal et al. 2018).

4.4 *Anthropocene Exhibit at the Art Gallery of Ontario*

From September 2018 to January 2019, the Art Gallery of Ontario in Toronto, Canada hosted "Anthropocene" a major contemporary art exhibit that illustrates how humans, individually and collectively, are leaving a signature on earth. Using film, photography and new experiential technologies world famous Canadian photographer Edward Burtynsky, and Canadian film makers Jennifer Baichwal and Nicholas de Pencier use the aesthetic impact of their imagery to trigger an important dialogue. They state: "Our ambition is for the work to be revelatory, not accusatory, as we examine human influences on the Earth both on a planetary scale and in geological time. The shifting of consciousness is the beginning of change." We have reached an unprecedented moment in planetary history where humans now change the Earth's systems more than all natural forces combined. This is the central argument that identifies the new geological epoch called The Anthropocene. They travelled to countries on every continent, except Antarctica, documenting irreversible marks of humans' impacts on the environment inspired by a desire to bear witness to the scale and gravity of humans' impact on the planet (AGO 2018).

From concrete seawalls in China that now cover 60% of the mainland coast, to the biggest terrestrial machines ever built in Germany, to psychedelic potash mines in Russia's Ural Mountains, to metal festivals in the closed city of Norilsk, to the devastated Great Barrier Reef in Australia and surreal lithium evaporation ponds in the Atacama desert, the filmmakers have traversed the globe using high end production values and state of the art camera techniques to document evidence and the experience of human planetary domination.

4.5 *Greta Thunberg and the Youth Movement Fighting Climate Crisis in 2019*

In her article "The Adults Have Failed Us" in Broadview, Alanna Mitchell discusses how youth led by Swedish teenager Greta Thunberg have in the past year taken the lead in fighting the climate crisis. In August 2018, when she was 15, Greta began staging her own school strikes to get her message to the Swedish government of the need to cut carbon emissions. In November, Greta created a TED talk in Stockholm to reach out to more than 2 million viewers in 30 languages with her carbon neutral message. She continues her activist work taking low level carbon imprint transportation (trains) in Europe and a sailboat to North America to speak at the United Nations. The student demonstrations continue worldwide and Greta has become a celebrity for her climate crisis cause. Legislatures in many countries, including Canada's House of Commons, are continuing Greta's work (Mitchell 2019).

5 Conclusion: Is Wilderness Protection Still Possible?

Is protection of the wilderness just an aspiration or is it an achievable goal? Today it is very clear that the preservation of wilderness, and particularly forests, is one essential means for sequestering carbon to address climate change caused by man-made green-house gas emissions. Further, the reduction in green-house gas emissions is a means to a healthy environment which is regarded as a universal right under the UN Charter of Rights and Freedoms and, by definition, a right that applies to the majority of citizens in democratic states.

Therefore, all democratic states should have constitutions that include the right to a healthy environment. Among the responsibilities for realizing this right is the protection and expansion of wilderness areas for addressing climate change. Since 1970, 130 countries have amended their constitutions to include a state obligation to protect the environment or the right to a safe, healthy and ecologically sustainable environment (Brown and Brown 2015, pp. 170–171). Only a few have specific constitutional provisions for protecting wilderness areas as I have noted above.

So the answer to the question depends upon constitutional rights which shifts the balance of political power to the judiciary. As Boyd observes, the presence of rights to a healthy environment in constitutions represents a remarkable evolution of the judiciary whose focus for centuries has been limited to the protection of private interests as opposed to public interests such as stewardship over the land and ecological systems of the nation state.

However, the fact remains that the 20 nations of the world with exclusive common-law systems have no express environmental rights of a healthy environment in their constitutions. As a consequence, they lag behind the other nations in recognizing the right to a healthy environment. Boyd asks the question "Is their objection based on the issue of justiciability or do they block progress towards social, economic and cultural rights as well as rights to a healthy environment because of fears related to the distribution of power and resources (Boyd 2012, p. 288)?"

The answer to the last of these question is a likely yes. In his 2018 book *Can Democracy Survive Global Capitalism?* Robert Kuttner explains how the politics of a succession of US Presidents and Congresses since 1973 have changed a decent democratic social contract in the US that developed during the three decades following World War II and fostered a huge increase in the concentration of wealth in the top ten percent of its citizens today (Kuttner 2018). This outcome signals the end of democracy as defined by former supreme court justice Louis Brandeis who famously stated in the 1940s "We can have democracy or we can have wealth concentrated in the hands of a few but we cannot have both" (Freeland 2012, p. 277).

New and amended constitutions show that politics can impact on environmental provisions in constitutions with positive or negative results. The large number of amended constitutions with environmental provisions since 1970 and the major increase in public awareness of climate change around the globe in the last ten years indicate that there is some room for optimism.

References

AGO (2018) Anthropocene Exhibit. https://www.AGO.ca/exhibitions/anthropocene. Accessed 15 May 2019

Aulakh R (2014) Tunisia's next Arab Spring: fighting global warming. Toronto Star. 2 Feb

Baichwal J, de Pencier N, Burtynsky E (2018) Anthropocene: the Human Epoch. https://theanthropocene.org/film. Accessed 15 May 2019

Banda M (2018) Why should trees have legal rights? It's second nature. The Globe and Mail. 2 June, p O5

Bergandi D (2017) The ecological catastrophe: the political-economic caste as the origin and cause of environmental destruction and the pre-announced democratic disaster. In: Westra L, Gray J, Gottwald F-T (eds) The role of integrity in the governance of the commons: governance, ecology, law, ethics. Springer, pp 179–189

Boyd D (2012) The environmental rights revolution: a global study of constitutions, human rights, and the environment. UBC Press, Vancouver

Brown B, Brown D (2015) Commonly unrecognized benefits of a human rights approach to climate change. In: Westra L, Gray J, Karageorgou V (eds) Ecological systems integrity: governance, law and human rights. Routledge, New York, pp 162–175

Brown C (2014) Kayapo Courage. https://www.nationalgeographic.com/magazine/2014/01/kayapo-courage. Accessed 20 May 2019

Carson R (1962) The silent spring. Houghton Mifflin, Boston

Freeland C (2012) Plutocrats: the rise of the new global super-rich and the fall of everyone else. Doubleday, Canada

Government of Canada (1982) Canadian Charter of Rights and Freedoms

Kitchell M (2014) A Fierce Green Fire. https://www.pbs.org/wnet/americanmasters/a-fierce-green-fire-about-the-film/2924. Accessed 8 Oct 2019

Kuttner R (2018) Can democracy survive global capitalism? W.W. Norton, New York

Mauro I, Suzuki D (2018) Beyond Climate. https://www.beyondclimate.ca/film. Accessed 22 Oct 2018

Mitchell A (2019) The adults have failed us. Broadview, Observer Publications, Toronto
Reed S, Castle S, Eddy M (2014) Europe dilutes clean-energy ambition. The Toronto Star. New York Times International Weekly. 26 January 2014
Tabachnick D, Koivukoski T (eds) (2011) On oligarchy: ancient lessons for global politics. University of Toronto Press, Toronto
Wikipedia, John Muir. https://en.wikipedia.org/wiki/Joh-Muir. Accessed 15 May 2019
Wikipedia, Wilderness. https://en.wikipedia.org/wiki/Wilderness. Accessed 15 May 2019

Quantum Mechanics and Law: What Does Quantum Mechanics Teach Us?

Orsolya Bányai

Abstract In this chapter the modern natural sciences, especially the findings of the subjective interpretation of quantum mechanics will be integrated into legal theory. The chapter builds on the thesis that ecological problems occur because man-made law is not consistent with the fundamental laws and principles that govern how the Universe functions. According to the author, these higher fundamental rules, similarly as in the legal theory of St. Aquinas, are different in their nature. Therefore, the study distinguishes between eternal and universal laws. Eternal laws are transcendental in their nature and universal laws are immanent. In order to handle ecological crisis in an effective way, these higher rules should be acknowledged and followed by human law. The meaning and the content of eternal and universal truths is also discussed here. This idea presented in this chapter is underpinned by the theory of Pitrim Sorokin on social and cultural dynamics, as the presented worldview fits in Sorokin's Idealistic/Integrative Truth. Furthermore, as an 'integrative' worldview, it also fits into the interdisciplinary Integrative Worldview Framework (IWF). Finally, the main points of this chapter are assimilated in the findings of contemporary environmental law literature.

1 Environmental Crisis in a Nutshell

Our planet and our society are experiencing an ecological crisis. According to the scientific community, our society, by overstepping certain ecological boundaries, impaired the ecosystems of the Earth so much that the planet will not able to ensure safe living conditions for humanity. A few of these planetary boundaries (e.g., climate change, biodiversity loss) have already been violated and others (e.g., the decline of natural resources as metals, freshwater, forests) will be transgressed soon.

O. Bányai (✉)
University of Debrecen Faculty of Law, Environmental and Labour Law Department, Debrecen, Hungary
e-mail: banyai.orsolya@law.unideb.hu

L. Westra et al. (eds.), *Ecological Integrity in Science and Law*,
https://doi.org/10.1007/978-3-030-46259-8_13

This fact indicates to us that something is not in order and also indicates that environmental regulation is failed (Laitos and Okulski 2017; Fodor 2003). But what are the causes of the failure? According to the literature, there are many causes, on different levels: technical, ethical, legal, ontological. Technically the crisis is because of economic growth, or overconsumption (Borgman 1984). But besides this technical problem a lot of authors emphasize the ethical root of the problem as well. From an ethical perspective the main cause of the crisis is that anthropocentric worldview that distinguishes humans from the natural environment (duality) and supports the idea that humans can rule over it (domination). But mainly the deepest root is ontological, namely we do not know who we are. This perspective is represented not only by spiritual leaders, but it has a growing base in the philosophical and economic literature as well (Bányai 2019).

2 Legal Reasons Behind the Crisis

The crisis also has legal reasons. Besides inconsistency within the goals of legal regulation and its inflexibility to comply with anything to ensure ecological sustainability (Westerlund 2008), the ecological problem occurs mainly because man-made law is not consistent with the fundamental laws and principles that govern how the Universe functions (Cullinan 2011). In order to restore the harmony and health in both society and nature, human-made law should find, reconnect, and represent the fundamental truths of higher laws (Burdon 2015; Bosselmann 1995, 2013; Cullinan 2011; Laitos and Okulski 2017; Bányai 2019). However, the positivist approach to law denies everything higher than positive law, however the fact of the environmental crisis questions this attitude. This is because man-made legal systems fit in the whole ecosystem as dying organs. The legal environment is so misshapen and unhealthy, that it in itself blocks the overall and deep changes in our society which would be needed to adapt to ecological boundaries. If we want to heal the legal regime, by the acceptance of higher rules, it should be reconnected to the whole ecosystem. However, the question immediately arises as to what these higher rules/laws are? Firstly, the main contemporary and traditional theories will be presented below with regards to this question. Secondly, by highlighting the incoherency of these theories as a whole, a possible integrated solution will be recommended.

2.1 Eternal, Universal Laws: Contemporary Standpoints

In regard to the content of higher rules in the modern environmental law literature, there are different viewpoints. For example Cormac Cullinan in the 'Wild law' describes them as *"laws or principles that govern how the universe functions"*. He noted that they are *"timeless and unified in the sense that they all have the same source." It can be witnessed in the "phenomenon of gravity," "the alignment of the planets [...]"* (Cullinan 2003, p. 84). Peter Burdon provides more information in connection to what universal law means. According to him universal principles (*called Great Jurisprudence or Great Law*) that have to be obeyed by human law can be known from the science of ecology and measured with respect to concepts such as ecological integrity (Burdon 2013, p. 86) But the most elaborated answer comes from Jan Laitos and Juliana Okulski. In their book the symmetry is that universal principle that has to be followed by positive law. Symmetry means "*change without change*" and it contains three further principles such as conservation, equivalence, and unification (Laitos and Okulski 2017).

The common feature of these standpoints is that they try to understand universal laws with the help of science. Also another common characteristic of the previously mentioned positions is that all of them have an immanent approach.[1] Finally and most obviously the latest common element of these viewpoints is that all of them acknowledge the existence of a higher order, to which man-made law has to obey.

Such duality of human law and higher rules is a tradition in both western and eastern law and legal theory as well. For instance, between the above viewpoints of modern environmental laws scholars and natural law there is an inevitable similarity. This is recognized also by Peter Burdon, who tries to distinguish the Great Jurisprudence from Natural law (Burdon 2011a). In the theories of modern natural law proponents (e.g. John Finnis, Robert P. George) the focus is mainly on human reason[2] (Lloyd 2016). Nevertheless, Burdon states that the focus of the Great Jurisprudence is not on human reason as we can see in rationalist natural law approach, but on the principle of human-Earth community, supported with scientific evidence (Burdon 2011). This is so because '*certainly, scientific principles such as the concept of Earth community are much better suited to shaping human law than non-verifiable appeals to human reason as promoted in Natural Law*' (Burdon 2011, p. 149). Thus on the content of the Great Law the science provides approximate descriptions that are interpreted and applied by human lawmakers. Furthermore, in contrast to Natural Law, the Great Law is concerned with the physical environment; therefore its perspective is eco-centric not anthropocentric.

2.2 Eternal, Universal Laws: Traditional Standpoints

At this point we should turn back to the question as to what constitutes higher laws to which man-made law should obey? Traditional natural law theorists, in contrast to contemporary natural law philosophers, had completely different standpoints in this regard. For example, in the philosophy of Plato above human law ideal law exists. That is transcendental, which is beyond the rationally discernible. Only in a state of religious ecstasy can it be glimpsed by a few who have been selected by

[1] The term immanent is used here in the sense as within the space-time continuum. The meaning of this will be clearer later in this chapter.

[2] Note that natural law theorists recently show openness to exploring the role of the virtues and the emotions beside reason.

divine grace (Kelsen 1960). In the Summa Theologica, Aquinas distinguishes between eternal, natural, divine and human law. Eternal law is the absolute foundation of any other law because the natural law is a participation of the rational creature in the eternal law (Baur 2012),[3] and the positive laws at least cannot be in contradiction with the eternal law. According to him, '*the eternal law is nothing else than the type of Divine Wisdom, as directing all actions and movements*.' (Aquinas Question 93 Article 1) *No one can know the eternal law, as it is in itself, except the blessed who see God in His Essence. But every rational creature knows it in its reflection, greater or less*" (Aquinas II. Part Question 93 Article 2.) St. Augustine also shows that above human law there is an 'eternal and unchangeable' law, '*the law according to which it is just that all things be perfectly ordered*' (Augustine 1993). Regarding Eastern legal systems, for instance, the higher order-human law (*Rta-dharma*) duality also has a tradition in the classical Hindu law. In classical Hindu law eternal law called *Rta*, which refers both to the true and everlasting movements of the cosmos (Davis 2010). Rta evolved from the *shruti* (the Vedas,[4] divine revelations) (Timm 1992) and *smrti* (what was remembered). The *dharma* is the *Rta* applied to human field, it is a collection of concrete model of behaviors (Hacker 2006).

One common feature of these traditional standpoints is that to determine eternal law, all of them build on metaphysical roots. It means that the representatives of classical natural law had a deeper approach on rules above human law than most authors have in modern literature.[5] According to historian Heinrich Rommen in the history of natural law Grotius was the turning point. Hugo Grotius may be said to have marked the transition from the metaphysical to the rationalist natural law (Rommen 1998). Another common characteristic of the abovementioned natural law approaches is that eternal law can be known generally by revelation (Finnis 1980),[6] not reason or science as we could see earlier in this chapter. Finally, modern natural law is interpreted in an antrophocentric fashion, while the theories of classical natural law proponents can be considered rather theocentric. Regarding Plato and Aristotle John Finnis notes that they do not use the existence of God or the gods as an argument to justify their claim that there are objective norms of human flourishing and principles of human reasonableness. But their arguments in justifying that claim, and their reflection upon the nature, point, and source of those arguments, lead them to affirm that there is a transcendent source of being (Finnis 1980).

[3] 'It is important to note that, for Aquinas, natural law is not something separate from eternal law. Rather, for Aquinas, the natural law is the eternal law itself, but regarded under the aspect of its being in us (rational beings) in this unique, twofold way: it is as in created beings that are ruled, measured, and directed by means of it, but also in us as in created (rational) beings that rule, measure, and direct by means of it.'

[4] The Vedas are considered the most sacred books of India. They are the original scriptures of Hindu teachings, containing spiritual knowledge encompassing all aspects of life. From the Vedas in this study I will refer to the Bhagavad-Gíta ("Song of the Lord"; BG) This is one of the most popular and authentic Hindu texts and has an important place within the Hindu tradition.

[5] Note that this claim is true regarding the secular and not the religious reconsiderations of natural law.

[6] According to Aristotle the supreme object of understanding is God and that intelligence (nous) understands itself through participation in the object of understanding.

The basic difference between modern and historic positions on rules above human law has become evident by now: the contemporary authors *basically* have an immanent approach, while the traditional authors had a transcendental approach to higher laws. Furthermore, it is also important that different type of superior rules (eternal law, universal principles, natural law) can be discerned. Therefore, in the following when I talk about eternal law it is considered transcendental in its nature, when I refer to universal law/principle or natural law, I consider them to be immanent in their nature.

2.3 Creating One from the Two Standpoints

As was shown, with regards to the content of higher rules, the modern environmental law standpoints and classical natural law approaches do not match each other perfectly. Only traditional authors were talking about transcendental eternal laws. What happened with these laws? Two answers are possible here. First, traditional theorists were wrong and there are no eternal laws above human law at all. Second, there are eternal laws, but most contemporary scholars ignore them. I'll argue here that transcendental eternal laws exist. Furthermore, I'll also hold that beside transcendental laws, there are also immanent rules above man-made laws. Thus, my opinion is that the reality in this question is not 'or-or', but 'so-so'. To support this opinion, in the next section I'll integrate the findings of several natural sciences. Finally I will also refer to the theory of the sociologist Pitrim Sorokin.

2.4 Bridging Science and Revelation

In order to show that there are eternal laws above human law, first of all the basic characteristic of such a legal order should be cleared. If we are just thinking logically, there are at least three requirements which help us to determine what eternal laws should look like (see Fig. 1).

1. They have to be independent from time. It means that eternal laws should be the same now, 5000 years ago or 100,000 years into the future. Additionally, Augustine argues that eternal law can never be altered by any circumstance that is unchangeable law.
2. Eternal law should be independent from space, which means that these rules have to be the same everywhere on the Earth, or even in universe(s).
3. Eternal laws have to be independent from the way they are known. It means that both science and revelation have to lead us, at least to a certain level, to the same eternal and universal truths.

Based on this presumption the reasoning can be continued by referring to the findings of natural sciences. In natural sciences there were radical improvements in the last hundred years; even those cannot be considered as mainstream ideas. According to a new scientific wave nowadays a paradigm-shift is occurring (Dossey

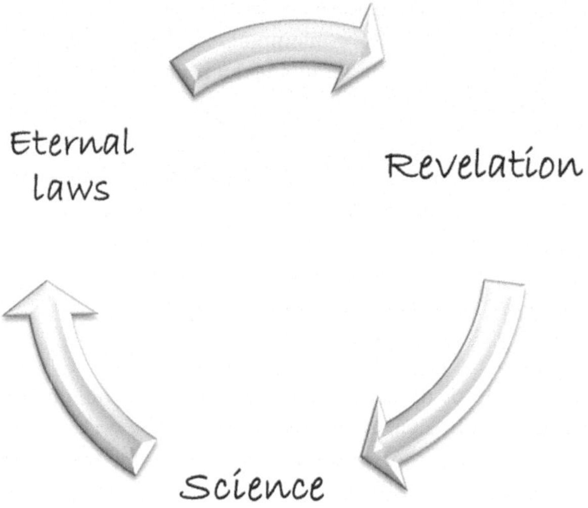

Fig. 1 Both science and revelations have to lead us to eternal truths

2017, p. 233; László 2014, p. 16; Bányai 2019). This is because different scientific disciplines, such as quantum physics, cosmology, nonlocal consciousness research, psychology, biology, have come to similar conclusions. The results of this new scientific wave suggest *that consciousness is fundamental, it is transcendental, which means that it is beyond the space-time continuum, and reality is not independent from consciousness.* This new paradigm comes from that special quantum physics experiments, as double slit experiment, the observer effect and the entanglement phenomenon. Furthermore, nonlocal consciousness researches also have a lot of empirical evidence on out of body experiences, near death experiences and extrasensory perceptions which underpin this worldview. It should be taken into account that nonlocal consciousness research, thanks to those unusual strict research protocols, works with 1 in a billion possibility of error. In order to understand what this means, we should consider the following: when the Higgs boson was found in the European Organization for Nuclear Research (CERN) in 2012, one in 300 million was the chance that the Higgs boson does not exist (Bányai 2019). Finally, it also must be emphasized that this 'new' paradigm is supported by numerous widely recognized members of scientific community (Bányai 2019).

Furthermore, this worldview which holds that consciousness is transcendental is in line with the scriptures as well. This is because the essential and traditional message of all scriptures, as the Bible, or the Bhagavad-Gíta (BG) (Bhaktivedanta Swami 1972), is that we are eternal spirits (St. Teresa of Avila 1921, p. 119) and we belong to the spiritual world (see Fig. 2).

As for the bridge between science and scriptures, below is the mentioned common part, which I call the 'Eternal Order' (Bányai 2019). The 'Eternal Order' mainly is built on the results of quantum physics and non-local consciousness research, but at the same time is in line with the Bhagavad-Gíta (and Upanishads) and the Bible as well. Therefore the 'Eternal Order' is in harmony with the previously mentioned presumption, which requires that eternal truths be independent

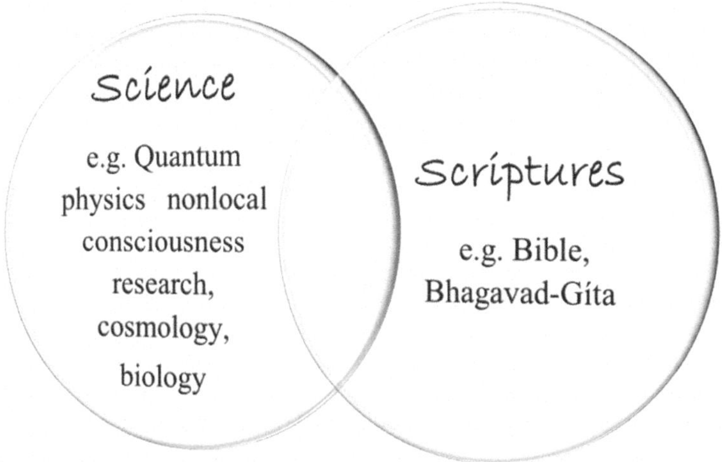

Fig. 2 The common part

from time, space and from the method by which they come to be known. It should be noted that this is not an exhaustive list; it can be expanded upon, which is symbolized by the fifth point here.

1. Consciousness of humans and that all living beings have a transcendental nature (*atman, spirit, pneuma*).[7]
2. All consciousness, regardless of its physical manifestations, is part of, and is interconnected with, a transcendental, universal Consciousness (*Paramatma, Holy Spirit, Pneuma*).[8] It must be emphasized that the universal whole is not the sum of the individual parts, similarly as in system theory. In this transcendental dimension, where the so called EPR-phenomenon occurs, the information transmission is immediate and mutual between individual and universal consciousness.
3. Individual consciousness depends on universal Consciousness by its position.
4. "Matter" depends on universal Consciousness by its nature. Matter here means everything which is not transcendental. Thus reality is the combination of matter (*prakriti*) and transcendental consciousness (*purusa*).[9]
5. …

[7] See: the theory of biocentrism; the Integrated Information Theory; BG 2.30; the Bible (Ecclesiastes 3:19) (King James Version): "For that which befalleth the sons of men befalleth beasts; even one thing befalleth them: as the one dieth, so dieth the other; yea, they have all one breath; so that a man hath no preeminence above a beast: for all is vanity"; the book of Enoch 59:1.; Furthermore according to the Christian mystics St. Teresa of Avila St. the soul and the spirit distinct though united.

[8] See: *BG* 13:17 and 13:28.

[9] See: *BG* 13:15 and 13:30.

My opinion is that the above mentioned scientific discoveries and the fact that these are in harmony with the scriptures verify both the existence and transcendental nature of a higher reality. This reality can be called the eternal order. It should be added that however this order is unchangeable it also has a dynamic aspect as our reality is always changing from moment to another.

The idea of the 'Eternal Order' fits in the theory of Pitrim Sorokin on Social and cultural dynamics (Matern 2017) as well. According to Sorokin in history there is a cyclical fluctuation between dominant social values, truths and ethics of the society. The three main systems of truth which fluctuate are the Ideational, the Sensational and Idealistic. In a purely ideational culture the dominant system of truth must be mainly "revelation" granted in a supersensory and even superlogical way, inspired by a superempirical agency or power, be it personal or impersonal. Based upon the revealed truth of God, absolute, perfect and omniscient, the truth is also believed to be absolutely certain in its validity (Sorokin 1937). However, in a Sensate society the truth based upon the testimony of the organs of senses has to be dominant. Since for the bearers of a Sensate mentality there is no reality behind the sensory reality, it is based upon scientific truth and knowledge. While the third Idealistic culture is one between the supersensory revelation and sensory evidence; one in which both previously mentioned systems are originally united. Here the value of the knowledge about the sensory phenomena is subordinated to that of the supersensory realities and revelations. In the era of the global ecological crisis we are living in a Sensational society, where the dominant characteristic of contemporary mentality has a utilitarian and hedonistic nature. Sorokin predicts that when there are no absolute moral standards—as in our age—, with further movement in this direction, the uncomfortable feeling is likely to increase until it reaches a stage when a shift to absolute moral standards becomes unavoidable (Sorokin 1937, p. 512). Thus Sorokin's theory also supports the necessity of acknowledgement of both higher rules above human law and those transcendental and immanent natures at the same time. Therefore, on the basis that there is a real common segment, where the laws of science and scriptures meet, I suppose that in the cycle of Sorokin our society is just commencing an Idealistic period (Sorokin 1937, p. 8).[10]

3 Conclusions

My opinion is that above human laws there are higher rules, and it should be acknowledged that man-made law has to obey higher rules. As for the nature of these higher rules, they are both transcendent and immanent. As in the theory of Aquinas, which is distinguished between eternal and natural law, the unchangeable space-time independent laws are eternal and transcendental, while those laws which are higher than

[10] 'Scholasticism' of Plato and Aristotle, of Albertus Magnus and St. Thomas Aquinas; or the 'scholasticism' of the Upanisadhs and other idealistic philosophies based upon the Vedas in India are examples of cultures which came close to "idealistic" or "integral" in the history.

man-made laws, but are not independent from space and time, such as the phenomenon of gravity, are immanent. Correspondingly, in Sorokin's Idealistic/Integrative society, both the sensory-empirical and the revealed truth get on well with each other. Therefore the concept of 'Eternal Order' presented previously, complements and partly covers the picture in the contemporary environmental law literature about higher laws. Partly covers the picture because its essence is similar to the principle of symmetry by Laitos. However, Laitos and Okulski explicitly do not mention the transcendental aspect of symmetry. It complements the picture, because it is a transcendental approach, while the concept of Peter Burdon and Cormac Cullinan is rather immanent. However, these immanent approaches are also useful and practical. Nevertheless the transcendental approach also should be utilized. This is because the results of modern science supports that our reality is subjective, as the "matter" depends on universal Consciousness. Therefore perceiving the reality through science is limited. It does not mean that it is useless, but its limits should be respected and the transcendental sphere also should be acknowledged even by the law.

The worldview of 'Eternal Order' handles the different causes of ecological crisis at the same time. Technically it serves as an excellent basis to revise modern positive law (E.g. constitutionalism; liability; private property, the rights of living entities, the right of the Earth; ecological sustainability) in order for legislation be able to give effective answers to the global ecological crisis. Furthermore, it solves the ethical problem of human-nature duality as well. This is because by the fact that all living entities are considered as a transcendental spirit, there is no more duality between man and other entities (Deák 2019). It also solves the problem of human domination, because the center is not the man but the universal and transcendental consciousness to whom we all belong (Quinn 2018). According to the Oslo manifestation the mind and brain duality is considered as a reason of our ecological crisis. I suppose *the above presented standpoint is not dual at all*, because as matter (*prakriti*) naturally follows the universal consciousness (*Purusa*) there is no gap between consciousness and matter. *Our reality is the absolutely indivisible complexity of matter and consciousness.* Finally, it also answers the ontological problem as who we really are. Thus, I can say that the acceptance of the worldview presented here might able to handle the root problems of today's ecological crisis in a deepest way. Or at least it is a first step in the recovering of society and nature.

I suppose that the above worldview fits in the category of Sorokin Idealistic/Integrative Truth. However, it also fits into the Integrative Worldview Framework (IWF) (Hedlund-de Witt 2012). IWF is an interdisciplinary framework that synthesizes research from a number of fields, notably developmental-structural psychology and sociology. The framework delineates between at least four major ideal-typical worldviews in the contemporary West: a traditional, modern, postmodern, and integrative. Within this frame the 'Eternal Order' and its legal considerations can be categorized as an 'integrative'[11] (Hedlund-de Witt and Hedlund-de

[11]The ontology of the 'integrated' worldview is that reality is transcendent, and immanent; the universe as evolving and creative manifestation of Spirit; nature as constructed and intrinsically valuable.

Witt 2013, p. 192) worldview. It can be said that generally no worldview is intrinsically better than another, however according to numerous research, the ‘integrative’ worldview is considered important in addressing ecological crisis, because *people with this worldview is inherently better motivated to behave in an environmental friendly way than people with ‘traditional’ or ‘modern’ worldview* (De Witt et al. 2016, p. 111; Cloninger 2013). Thus according to Annick de Witt policy-makers should pay careful attention to this worldview, using this knowledge to speak to it, draw support from it, market to it, and help create the conditions for its growth (De Witt et al. 2016, p. 111). My opinion is that without wider acceptance of this worldview, there is no hope for its notable academic, political, legal or juridical application. From this, a few challenging questions arise. For example, is there any room to apply legal tools by lawmakers to support an ‘integrative’ worldview within our current legal structures? If yes, what are these legally adequate solutions? If no, what other, non-legal tools are in order to promote it? Nevertheless, despite the hard task to support integrative worldview, another problem, which is considered usual in regard of natural law or even at Earth Jurisprudence as well, also occurs (Lloyd 2016). This problem concerns the *distance between theory and practice of the ‘Eternal Order’*. Or put in another way, how can the concept of ‘Eternal Order’ be converted into a practically applicable legal answer? Notwithstanding, the problem is not hopeless, because for instance the traditional Hindu law and other legal solutions from Sorokin’s idealistic societies might serve as a good starting point in order to implement theory in practice, at least at the level of academic reasoning.

References

Augustine (1993) On free choice of the will (trans: Villiams T). Hackett Publishing Company, Indianapolis

Bányai O (2019) The Foundation of an upcoming civilization able to reach its fulfillment within the ecological limits of the earth: the eternal order. World Fut 75:298–323

Baur M (2012) Law and natural law. In: Davies B, Stump E (eds) The Oxford Handbook of Aquinas. Oxford University Press, Oxford, pp 238–254

Bhaktivedanta Swami AC (1972) The Bhagavad-Gita as it is. Bhaktivedanta Book Trust

Borgman A (1984) Technology and the character of contemporary life: a philosophical inquiry. University of Chicago Press, Chicago

Bosselmann K (1995) When two worlds collide: society and ecology. RSVP, Memphis

Bosselmann K (2013) Grounding “The rule of law”. In: Voigt C (ed) Rule of law for nature: new dimensions and idea sin environmental law. Cambridge University Press, Cambridge

Burdon P (2011) Earth Jurisprudence: private property and earth community (Doctoral dissertation). Adelaide Law School, The University of Adelaide

Burdon P (2013) The Earth community and ecological jurisprudence. Oñati Socio-Legal Ser 5:815–837

Burdon P (2015) Earth jurisprudence: Private property and the environment. Routledge, New York

Cloninger CR (2013) What makes people healthy, happy, and fulfilled in the face of current world challenges? Mens Sana Monogr 11:16–24. https://doi.org/10.4103/0973-1229.109288

Cullinan C (2003) Wild law: a manifesto for Earth justice. Green Books, Dartington

Cullinan C (2011) Wild law. SiberInk, Cape Town

Davis DR Jr (2010) The spirit of Hindu law. Cambridge University Press, Cambridge
De Witt A, De Boer J, Hedlund N, Osseweijer P (2016) A new tool to map the major worldviews in the Netherlands and USA, and explore how they relate to climate change. Environ Sci Policy 63:101–112
Deák D (2019) Spirituality and law. In: Zsolnai L, Flanagan B (eds) The Routledge International Handbook of Spirituality in Society and the Professions. Routledge, New York, pp 280–287
Dossey L (2017) Consciousness and TDVP: welcome to a new world. Explore 13:233–237. https://doi.org/10.1016/j.explore.2017.04.001
Finnis J (1980) Natural law and natural rights. Clarendon Press, Oxford
Fodor L (2003) Milyen környezetjogunk legyen? [What our environmental law should be?]. Természet Világa 10:457–460
Hacker P (2006) Dharma in Hinduism. J Indian Philos 24:479–496
Hedlund-de Witt A (2012) Exploring worldviews and their relationships to sustainable lifestyles: towards a new conceptual and methodological approach. Ecol Econ 84:74–83
Hedlund-de Witt A, Hedlund-de Witt N (2013) The state of our world, the state of our worldview(s): the Integrative Worldview Framework as a tool for reflexive communicative action and transformation. In: University of Oslo (ed) Proceedings of Transformation in a changing climate. University of Oslo, Oslo, pp 187–202
Kelsen H (1960) Plato and the doctrine of natural law. Vanderbilt Law Rev 14:23–64
Laitos J, Okulski J (2017) Why environmental policies fail. Cambridge University Press, Cambridge
László E (2014) The science and Akashic field. An integral theory of everything. Inner Tradition, Rochester
Lloyd V (2016) New directions in natural law. J Law Religion 31:367–379. https://doi.org/10.1017/jlr.2016.37
Matern F (2017) Sorokin, Maritain and integral sociology. Études maritainiennes/Maritain Stud 33:116–129
Quinn N (2018) Applying the concept of ecological integrity in biosecurity law and management (Doctoral dissertation). Griffith Law School, Griffith University
Rommen HA (1998) The natural law: a study in legal and social history and philosopohy (trans: Hanley TR). Liberty Found, Indianapolis
Sorokin PA (1937) Social and cultural dynamics. American Book Company, New York
St. Teresa of Avila (1921) The interior castle (trans: Stanbrook B). Thomas Baker, London
Timm JR (1992) Texts in context: traditional hermeneutics in South Asia. State University of New York Press, Albany
Westerlund S (2008) Theory for sustainable development in sustainable development. In: Bugge HC (ed) International and national law. Europa Law Publishing, Amsterdam, pp 49–64

Part IV
Ecological Integrity and the Role of Earth Science

Evaluation of Ecological Integrity in Landscape Based on Remote Sensing Data

Jakub Zelený and Daniel Mercado-Bettín

Abstract Sustaining ecological integrity is recognized worldwide as a strategic objective (e.g. the 2015 Paris Agreement), but a general consensus on the overall methodology for assessing ecological integrity is still missing. This chapter presents a contribution to the method of ecological integrity evaluation, using simple and theoretically grounded method to calculate three holistic indicators: exergy capture, biotic water flows and abiotic heterogeneity, utilizing open access remote sensing data (Sentinel-2 and Landsat 8). Three variables are proposed as a representation of the respective indicators: NDVI (Normalized Difference Vegetation Index), brightness temperature (BT) and vegetation surface heterogeneity (HG). Forests and wetlands have obtained higher results in the selected integrity indicators, while arable lands and urban areas relatively lower. The relative distance between the potential peak and the lowest performance in a landscape context is obtained by calculating a composite Regional Index of Ecological Integrity (RIEI [%]). The proposed approach can be used for various purposes including localization of naturally valuable areas, estimation of ecosystem condition or performance of ecosystem management.

J. Zelený (✉)
Faculty of Humanities, Charles University in Prague, Prague, Czech Republic

D. Mercado-Bettín
Catalan Institute for Water Research, University of Girona, Girona, Spain

Grupo GIGA, Escuela Ambiental, Facultad de Ingeniería, Universidad de Antioquia, Medellín, Colombia

L. Westra et al. (eds.), *Ecological Integrity in Science and Law*,
https://doi.org/10.1007/978-3-030-46259-8_14

1 Introduction

1.1 Ecological Integrity and Self-Organization

Despite the universal uderstanding of the importance of maintaining ecological integrity (e.g. Paris Agreement 2015; Convention on Biological Diversity Aichi Targets 2019), a consensus on a unified theory or methodology of assessing integrity is still missing. Historically, integrity has has been understood as the degree of "naturalsness" or the distance from a natural reference (Karr and Dudley 1981; Majer and Beeston 1996; Scholes and Biggs 2005; Coppedge et al. 2006; Capmourteres and Anand 2016; for a review see Ruaro and Gubiani 2013). The approach based on naturalness estimation is known as "biotic integrity" and usually comprises biodiversity monitoring of selected sites or specific ecosystems such as forests, lakes or rivers (Karr and Dudley 1981; Fraser et al. 2009). Up till now, several regional scale frameworks have been proposed, adressing the multitude of ecosystem which landscapes consist of in a holistic manner, although a practical methodology was not empirically tested so far (Slocombe 1992; Andreasen et al. 2001; Reza and Abdullah 2011).

The aim of this chapter is to describe a novel method of ecosystem integrity evaluation in landscape context. The experimental definition of integrity used within this study is: "*Ecological integrity is the degree of self-organization. It is regulated by different constrains imposed by abiotic factors and human management*". This study utilizes the understanding of ecological integrity as the degree of self-organization (Müller et al. 2000; Müller 2005), or autopoiesis (Maturana and Varela 1980). Unlike the "biotic integrity" concept, the proposed approach does not quantify the distance to a natural reference, but rather the thermodynamic performance of an ecosystem (Schneider and Kay 1994; Maes et al. 2011) or the degree of self-organization.

Three indicators are used to estimate ecological integrity: exergy capture, biotic water flows and abiotic heterogeneity (after Müller 2005). Based on a preceeding literature survey, three variables were selected to represent these indicators: brightness temperature (BT), Normalized Difference Vegetation Index (NDVI) and vegetation surface heterogeneity (HG). The capacity of vegetation to reduce the temperature gradient is a promising measure of ecosytem metabolism and integrity, and was already proposed by Schneider and Kay (1994) and Maes et al. (2011). The amount of temperature reduced by ecosytems is directly linked to the volume of water transported through vegetation during evapotranspiration and can be thus linked with the ecological integrity indicator "Biotic water flows" proposed by Müller (2005). NDVI is a straightforward estimate of the amount of solar radiation absorbed during photosynthesis and can be thus linked with the indicator "Exergy capture" (Müller 2005; Kandziora et al. 2013). The third indicator, which is ecosystem "abiotic heterogeneity", can be estimated as the degree of ecosystem uneveness or complexity (Müller 2005; Parrott 2010), and stands in oposition to surface land surface homogeneity, which is typical for industrial land use.

This study explores whether the selected paratmeters, quantified using open-acess remote sensing data, can be used as a sufficient, flexible and comparable meassure of ecosystem integrity quantification. Part of the approach is testing an appropriate method of data aggreagtion, which would allow simple, yet informative landscape assessment, which is most desired in conditions of limited personal and material resources (CBD 2003).

2 Materials and Methods

2.1 Study Area

Protected Landscape Area (PLA) Třeboňsko is a site of significant natural value and a biospherical reserve in southern Bohemia (49°0′0,62″N, 14°50′7,79″E, approx. 700 km^2). It has been given the status of protected landscape area in 1979 and acknowledgend as a biospherical reserve in 1977. Třeboňsko is a lowland with over 400 lakes, primarily used for fish production. Besides fisheries, significant sectors comprise forestry and agriculture. Despite being a protected area, Třeboňsko is a completely man-made landscape created by draining extensive swamps and mires during the middle ages. Třeboňsko has therefore a significant natural as well as cultural and historical value.

2.2 Remote Sensing

Two sources of remote sensing data were utilized in the study: Landsat 8 (United States Geological Survey) and Sentinel-2 (European Space Agency). Landsat 8 provides multispectral as well as thermal measurements in 30 m resolution, while Sentinel-2 provides multispectral measurements only, but in considerably higher resolution (10–20 m). Both instruments provide comparable meaurements of vegetation photosynthetic performance (Roy et al. 2014; Castaldi et al. 2016; van der Werff and van der Meer 2016; Chrysafis et al. 2017).

Sentinel-2 (S-2) multi-spectral sensor is particularily suitable for assessing vegetation photosynthetic characteristics using the the so-called "vegetation indices" (Delegido et al. 2011; Krofcheck et al. 2016; Vincini et al. 2016). S-2 has been successfully used in agriculture (Vincini et al. 2016; Immitzer et al. 2016), forestry (Schumacher et al. 2016; Majasalmi and Rautiainen 2016), soil assessment (Poggio and Gimona 2016) and geology (van der Werff and van der Meer 2016; Castaldi et al. 2016; Forkuor et al. 2017).

Landsat 8 (L8) measurements have similar utility to S-2, but with lower spatial (30 m for multi-spectral and 100 m for thermal) as well as temporal resolution (there are two S-2 units—S2A and S2B with a 5 day revisit time at the equator; L8 has a 16 day revisit time). Unlike S-2, L8 carries a thermal sensor (TIRS - Thermal InfraRed Sensor) which is why the two satellites were combined to deliver a complementary set of vegetation measurements.

2.3 Data Sources and Processing

The three selected indicators of ecological integrity (exergy capture, biotic water flows and abiotic heterogenity) are represented by three respective variables: Normalized Difference Vegetation Index (NDVI), at-satellite brightness temperature (BT) and vegetation surface heterogeneity (HG). NDVI and HG was derived from S-2 images, while BT was obtained from L8.

2.3.1 Exergy Capture

The ecosystem exergy capture was represented using the variable NDVI. NDVI can be calculated according to Eq. (1),

$$NDVI[-] = \frac{NIR - RED}{NIR + RED} or \frac{Band8 - Band4}{Band8 + Band4}, \tag{1}$$

where NDVI is the Normalized Difference Vegetation Index, NIR is the reflectance in near infrared (S-2 Band 8) and RED means reflectance in red part of the visible spectrum (S-2 Band 4). It is a fairly straightforward estimate, closely related to primary production (Xu et al. 2012).

2.3.2 Biotic Water Flows

We have used at-satellite brightness temperature, obtained from L8 Band 10, as a representative of the indicator biotic water flows, following the Eq (2)

$$BT[^{\circ}C] = \frac{K_2}{\ln \frac{K_1}{L\lambda + 1}} - 273{,}15, \tag{2}$$

where BT is at-satellite brightness temperature, Lλ represents the parameter Top Of Atmosphere (TOA) spectral radiance [watts/(m^2 ∗ srad ∗ μm)] and K1 and K2 are both band-specific thermal conversion constants obtained from the meta-data of the image.

BT was further converted from an absolute temperature to a range in °C (upper and lower 2% were clipped to remove outliers) and inverted to represent the capacity of vegetation to reduce surface temperature, without the information on seasonal variation of temperature. The resulting variable is termed TD (Temperature Difference) and is calculated using Eq. (3)

$$TD\left[{}^{\circ}C\right] = (-BT) + \max BT, \tag{3}$$

where TD is the variable temperature difference, which is obtained from BT (at-satellite brightness temperature), calculated using Eq. (2)

2.3.3 Abiotic Heterogeneity

Vegetation surface heterogeneity (HG) was calculated using S-2 Band 4 and Band 8 and a DCED (Diagonal Compass Edge Detector) filter algorithm to quantifying geometric and linear shapes. The algorhitm is an integral part of the dedicated software for Sentinel missions—SNAP (ESA SNAP 2019). The respective calculation was done using Eq (4)

$$HG[-] = \left(\sqrt{|DCED(NIR)|}\right) - \left(\sqrt{|DCED(RED)|}\right), \tag{4}$$

which describes the calculation of the variable HG (vegetation surface heterogeneity). The DCED filter product from NIR (reflectance in near infrared, S-2 Band 8) and RED (reflectance red part of the spectrum, S-2 Band 4) was used to create a vegetation surface heterogeneity index by subtracting one product from the other. A near-normal data distribution was obtained by square root transformation, to enable comparison with the other two variables.

2.3.4 Data Sampling

The three variables were calculated for all available cloud-free images in the year 2017 and median values were sampled for individual land-use classes using the QGIS software (zonal statistics plugin). Variables were sampled using a land cover classification layer, which was specifically developed for the Czech republic (Consolidated Layer of Ecosystems—CLE, Frélichová et al. 2014).

2.3.5 Land-Use Integrity Estimation

In order to represent the temporal performance of each variable in the respective land-use classes, an Area Under Curve (AUC) was calculated, quantifying the volume of space of a hypothetical square between two points in time. This approach was selected as a simple method of data aggregation, representing the performance of each land-use class within the selected year (2017). Equation (5) describes the calculation of a single AUC between two measurements

$$AUC_t = \left(\frac{y_{t+1} - y_t}{2}\right) * \left(x_{t+1} - x_t\right), \tag{5}$$

where y represents the measured value in time t and x represents the number of days within a given year. An aggregated value for the whole year was calculated by adding each of the AUCs according to Eq. (6)

$$AUC = \sum_{t=0}^{n} AUC_t + AUC_{t+1} + \ldots + AUC_{t+n}, \tag{6}$$

where AUC is obtained by summing all of the indivitual AUC_t in the selected year 2017. The resulting values obtained for each land-use class were normalized (range 0–1, upper and lower 2% were set to maximum to remove outlyers) and expressed as percentage [0–100%]. The score 100% was therefore assigned to such land-use class, according to the Consolidated layer of ecosystems, which had the highest performance in the respective indicator.

2.3.6 Quantification of a Regional Integrity Index

An index of ecological integrity for each land-use class was obtained by averaging the three ecological integrity variables (mean EI, expressed as %) and multiplied by the fraction of area covered [0–1] to obtain an index of ecological integrity (IEI [%], Fig. 1) representing the proportional contribution of the respective land-use class to the overall integrity of the study area. Finally, an integrative Regional Index of Ecological Integrity (RIEI [%], Fig. 1) was calculated by adding the individual indexes (IEIs) for the individual land-use class.

	HG	NDVI	TD	EI	% Area	IEI
Beech forests	92%	99%	100%	97%	1.0%	1%
Floodland and wetland forests	100%	100%	90%	97%	1.9%	2%
Managed forests mixed	87%	98%	91%	92%	1.1%	1%
Natural bushes	94%	90%	80%	88%	1.2%	1%
Spruce forests	64%	98%	99%	87%	1.9%	2%
Managed forests coniferous	63%	99%	95%	86%	37.2%	32%
Oak and hornbeam forests	72%	97%	86%	85%	4.5%	4%
Peat forests	53%	96%	97%	82%	5.3%	4%
Managed forests broadleaved	93%	96%	55%	81%	0.2%	0%
Rocks Rubble	57%	93%	87%	79%	0.0%	0%
Wetlands and riparian vegetation	72%	79%	77%	76%	1.7%	1%
Dry pine forests	45%	94%	86%	75%	2.2%	2%
Sport and nleisure surfaces	58%	74%	89%	74%	0.2%	0%
Mesophilic meadows	53%	88%	65%	69%	2.0%	1%
Non-natural bushes	86%	90%	22%	66%	0.4%	0%
Urban greenery	74%	91%	30%	65%	0.1%	0%
Aluvial and wet meadows	44%	85%	43%	57%	3.0%	2%
Peat bogs and springs	60%	83%	9%	51%	0.2%	0%
Managed meadows	36%	86%	29%	50%	17.1%	9%
Continuous urban fabric	56%	53%	35%	48%	0.0%	0%
Dumps and construction sites	7%	27%	83%	39%	0.0%	0%
Traffic infrastructure	42%	39%	26%	36%	0.6%	0%
Discontinuous urban fabric	45%	49%	11%	35%	3.3%	1%
Arable land	7%	40%	38%	28%	14.2%	4%
Rocks and Quarrys	3%	21%	37%	20%	0.4%	0%
Industrial and business units	0%	0%	0%	0%	0.2%	0%
						67%

Fig. 1 A table presenting the relative performance of the tested parameters for individual land-use classes, the % of area covered, the average ecological integrity (EI) and integrity weighted by the % of area covered by a land-use class (IEI—Index of Ecological Integrity). A sum of the IEIs is used to calculate the final Index of Regional Ecological Integrity (RIEI). Variables representing EI indicators: HG—Heterogeneity, NDVI—Normalized Difference Vegetation Index, TD—Temperature Difference, were aggregated across the respective vegetation period (year 2017 in PLA Třeboňsko)

3 Results

3.1 Ecological Integrity of Variable Land-Use

The resulting performance in terms of ecological integrity has been tested and validated using a land-cover classification layer (CLE). Generally speaking, natural and managed forests, followed by wetlands had the highest performance in all three parameters. On the other hand, arable lands and urban areas (e.g. mineral extraction sites, sparsely vegetated areas) received the lowest values in all three EI variables (see Fig. 1).

3.2 Comparison of the Selected Variables

From a general perspective, forests, wetlands and grasslands have received the highest values in all of the three selected variables, while sparsely vegetated or urban areas have received te lowest. Figure 2 presents the composite result of the three variables, obtained by normalization and averaging. This method can be used to evaluate and detect areas of high ecosystem integrity. Areas which have received the highest values are already acknowledged under some form of protection (AOPK 2019), or comprise highly productive forests (personal communication with representatives of the nature conservation agency). A focus on a highly valuable wetland known as "Novořecké močály" is included in the figure to demonstrate the performance of the composite EI index in detail. The subject wetland vegetation has received an overall highest EI score, while the nearby clearcuts (red geometric objects) have received relatively low score. This example demonstrates the capacity of the three variables and their composite index to produce a gradient between highly self-organized and disturbed land cover.

Table 1 presents the results of a Spearman's correlation test to demonstrate the relationship between the individual variables. NDVI and TD were the most tightly correlated pair, while TD and HG were the least correlated.

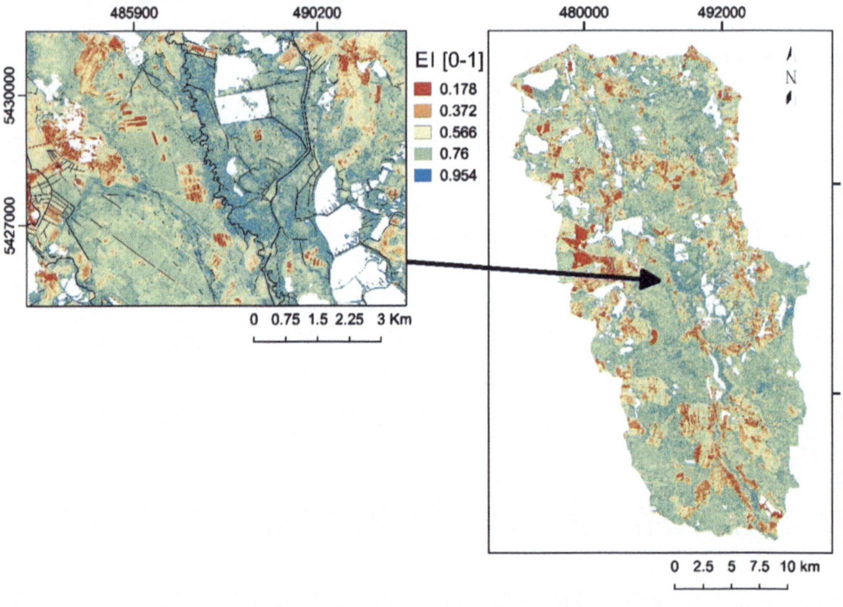

Fig. 2 A product of normalization and averaging of the three selected variables (NDVI, TD, HG) in the whole study area PLA Třeboňsko and a focus of a selected site of interest—a highly valuable wetland "Novořecké močály." The images used were sensed on 13.6.2017 (S-2) and 20.6.2017 (L8) and produced in 10 m resolution. Blue colour represents high integrity (dense vegetation, forest cover) while red highlights low integrity areas (urban, disturbed or harvested ecosystem). EI stands for ecological integrity

Table 1 Results of a correlation test between the four variables, using the Spearman's correlation method

	NDVI	TD	HG
NDVI	*	0.47	0.41
TD	0.47	*	0.29
HG	0.41	0.29	*

All correlations were significant (p-value < $2.2{*}e^{-16}$). Images analysed were sensed on 13.6. (S-2; HG, NDVI) and 20.6. 2017 (L8; TD)

4 Discussion

4.1 Indicator Design

The aim of this study was to design and validate suitable variables to represent a set of selected ecological integrity indicators. Second part of the study was the design of an appropriate aggregation method with the aim to provide a relative quantification of individual land-use types as well as the overall performance of the selected region. Although the three tested variables are not new in landscape assessment (Schneider and Kay 1994; Xu et al. 2012; Parrott 2010; Maes et al. 2011; Vargas et al. 2017), the approach to use them in combination is novel.

The correlation test in Table 1. indicates, that the selected variables are related and we assume they describe three different aspects of a single phenomenon, which is self-organization. The initial goal of the selected variables was to provide a means of differentiation between anthropogenic land-use and natural ecosystems and this was achieved. Additional parameters can be used to further supplement the current set. These comprise for e.g. vegetation respiration, gross and net primary production, biotic heterogeneity (biodiversity), standing biomass and ecosystem nutrient accumulation capacity, although more potential variables exist (Müller 2005; Rocchini et al. 2018).

4.2 Limitations of the Method

The presented approach to produce is based on several assumptions. Firstly, the indexes for individual land-use classes represent an interval, not an absolute value. Their relative performance in the context of the selected case sutudy area is determined by the respective regional highest and lowest performing land-use type, serving as a reference for both extremes. This means, that the result is always dependent on an appropriate site selection method, which should be large enough to contain sufficient anthropogenic as well as natural references (municipalities, countries, federal states). An obvious objection to the approach would be, that the results will always be dependent on the specific site selection and are thus not fully comparable among different case study areas. To defend the presented method, it can be said that in the first instance, the variables are usable to detect ecological integrity even without normalization. Normalizations within specific land-use classes have been used mainly to validate the variables with respect to their capacity to differentiate natural from anthropogenic land use types. The relative indexes, including the resulting RIEI, are suitable for comparison between different regions; their interpretation has to take into consideration the specifics and limitations of the aggregation method.

5 Conclusions

Three variables representing three ecological integrity indicators have been designed based on open-acess remote sensing data and validated using a land-cover layer (CLE). Valuable areas from the point of view of nature conservation as well as ecosystem management can be highlighted using the proposed variables, along with areas of low value and disturbed areas. A method of aggregation is proposed, quantifying the relative distance of a land-use class from a regional maximum and minimum to provide a method of land-use integrity evaluation and quantification. A composite Regional Index of Ecological Integrity (RIEI) has been designed and quantified to provide an estimation of the overall regional integrity and it's distance from a hypothetical top and bottom state.

Acknowledgements We would especially like to thank our colleagues from the Institute of Natural Resource Conservation of the University in Kiel, namely Felix Müller, Wilhelm Windhorst and Claus-G. Schimming for consultation and advices on the methodological aspects and results interpretation. This study was made possible thanks to financial support from the DBU (Deutsche Bundesstiftung Umwelt) MOE scholarship, GAUK (Charles University Grant Agency [Grant no. 546517]) and Specific University Research [VS 260471] funding.

References

Andreasen JK, O'Neill RV, Noss R, Slosser NC (2001) Considerations for the development of a terrestrial index of ecological integrity. Ecol Indic 1:21–35. https://doi.org/10.1016/S1470-160X(01)00007-3

Capmourteres V, Anand M (2016) Assessing ecological integrity: a multi-scale structural and functional approach using structural equation modeling. Ecol Indic 71:258–269. https://doi.org/10.1016/j.ecolind.2016.07.006

Castaldi F, Palombo A, Santini F, Pascucci S, Pignatti S, Casa R (2016) Evaluation of the potential of the current and forthcoming multispectral and hyperspectral imagers to estimate soil texture and organic carbon. Remote Sens Environ 179:54–65. https://doi.org/10.1016/j.rse.2016.03.025

Chrysafis I, Mallinis G, Siachalou S, Patias P (2017) Assessing the relationships between growing stock volume and Sentinel-2 imagery in a Mediterranean forest ecosystem. Remote Sens Lett. 8(6):508–517. https://doi.org/10.1080/2150704X.2017.1295479

Coppedge BR, Engle DM, Masters RE, Gregory MS (2006) Development of a grassland integrity index based on breeding bird assemblages. Environ Monit Assess 118(1–3):125–145. https://doi.org/10.1007/s10661-006-1237-8

Delegido J, Verrelst J, Alonso L, Moreno J (2011) Evaluation of sentinel-2 red-edge bands for empirical estimation of green LAI and chlorophyll content. Sensors 11(7):7063–7081. https://doi.org/10.3390/s110707063

Forkuor G, Hounkpatin OKL, Welp G, Thiel M (2017) High resolution mapping of soil properties using remote sensing variables in South-Western Burkina Faso: a comparison of machine learning and multiple linear regression models. PLoS One 12(1):e0170478. https://doi.org/10.1371/journal.pone.0170478

Fraser RH, Olthof I, Pouliot D (2009) Monitoring land cover change and ecological integrity in Canada's national parks. Remote Sens Environ 113(7):1397–1409. https://doi.org/10.1016/j.rse.2008.06.019

Frélichová J, Vačkář D, Pártl A, Loučková B, Harmáčková ZV, Eliška L (2014) Integrated assessment of ecosystem services in the Czech Republic. Ecosyst Serv. https://doi.org/10.1016/j.ecoser.2014.03.001

Immitzer M, Vuolo F, Atzberger C (2016) First experience with Sentinel-2 data for crop and tree species classifications in central Europe. Remote Sens 8(3). https://doi.org/10.3390/rs8030166

Kandziora M, Burkhard B, Müller F (2013) Interactions of ecosystem properties, ecosystem integrity and ecosystem service indicators—a theoretical matrix exercise. Ecol Indic 28:54–78. https://doi.org/10.1016/j.ecolind.2012.09.006

Karr JR, Dudley DR (1981) Ecological perspective on water quality goals. Environ Manage 5(1):55–68. https://doi.org/10.1007/BF01866609

Krofcheck DJ, Eitel JUH, Lippitt CD, Vierling LA, Schulthess U, Litvak ME (2016) Remote sensing based simple models of GPP in both disturbed and undisturbed piñon-juniper woodlands in the southwestern U.S. Remote Sens 8(1). https://doi.org/10.3390/rs8010020

Maes WH, Pashuysen T, Trabucco A, Veroustraete F, Muys B (2011) Does energy dissipation increase with ecosystem succession? Testing the ecosystem exergy theory combining theoretical simulations and thermal remote sensing observations. Ecol Modell 222(23–24):3917–3941. https://doi.org/10.1016/j.ecolmodel.2011.08.028

Majasalmi T, Rautiainen M (2016) The potential of Sentinel-2 data for estimating biophysical variables in a boreal forest: a simulation study. Remote Sens Lett 7(5):427–436

Majer JD, Beeston G (1996) The biodiversity integrity index: an illustration using ants in Western Australia. Conserv Biol 10(1):65–73. https://doi.org/10.1046/j.1523-1739.1996.10010065.x

Maturana HR, Varela FJ (1980) Autopoesis and cognition. the realization of the living., Maquinas y Seres vivos, Boston stu, 59–138, doi:EB N MATU

Müller F (2005) Indicating ecosystem and landscape organisation. Ecol Indic 5(4):280–294. https://doi.org/10.1016/j.ecolind.2005.03.017

Müller F, Hoffmann-Kroll R, Wiggering H (2000) Indicating ecosystem integrity - theoretical concepts and environmental requirements. Ecol Modell 130(1–3):13–23. https://doi.org/10.1016/S0304-3800(00)00210-6

Paris Agreement (2015) European Commission's Directorate-General for Climate Action, 1–16. Available at https://ec.europa.eu/clima/policies/international/negotiations/paris_en. Accessed 30 May 2019

Parrott L (2010) Measuring ecological complexity. Ecol Indic 10(6):1069–1076. https://doi.org/10.1016/j.ecolind.2010.03.014

Poggio L, Gimona A (2016) Assimilation of optical and radar remote sensing data in 3D mapping of soil properties over large areas. Sci Total Environ 579:1094–1110. https://doi.org/10.1016/j.scitotenv.2016.11.078

Reza MIH, Abdullah SA (2011) Regional Index of Ecological Integrity: a need for sustainable management of natural resources. Ecol Indic 11(2):220–229. https://doi.org/10.1016/j.ecolind.2010.08.010

Rocchini D, Luque S, Pettorelli N, Bastin L, Doktor D, Faedi N, Feilhauer H, Féret JB, Foody GM, Gavish Y, Godinho S, Kunin WE, Lausch A, Leitão PJ, Marcantonio M, Neteler M, Ricotta C, Schmidtlein S, Vihervaara P, Wegmann M, Nagendra H (2018) Measuring β-diversity by remote sensing: a challenge for biodiversity monitoring. Methods Ecol Evol 9(8):1787–1798. https://doi.org/10.1111/2041-210X.12941

Roy DP, Wulder MA, Loveland TR, Woodcock CE, Allen RG, Anderson MC, Helder D, Irons JR, Johnson DM, Kennedy R, Scambos TA, Schaaf CB, Schott JR, Sheng Y, Vermote EF, Belward AS, Bindschadler R, Cohen WB, Gao F, Hipple JD, Hostert P, Huntington J, Justice CO, Kilic A, Kovalskyy V, Lee ZP, Lymburner L, Masek JG, McCorkel J, Shuai Y, Trezza R, Vogelmann J, Wynne RH, Zhu Z (2014) Landsat-8: science and product vision for terrestrial global change research. Remote Sens Environ 145:154–172. https://doi.org/10.1016/j.rse.2014.02.001

Ruaro R, Gubiani ÉA (2013) A scientometric assessment of 30 years of the Index of Biotic Integrity in aquatic ecosystems: applications and main flaws. Ecol Indic 29:105–110. https://doi.org/10.1016/j.ecolind.2018.12.016

Schneider ED, Kay JJ (1994) Life as a manifestation of the second law of thermodynamics. Math Comput Model 19(6):25–48. https://doi.org/10.1016/0895-7177(94)90188-0

Scholes RJ, Biggs R (2005) A biodiversity intactness index. Nature 434(7029):45–49. https://doi.org/10.1038/nature03289

Schumacher P, Mislimshoeva B, Brenning A, Zandler H, Brandt M, Samimi C, Koellner T (2016) Do red edge and texture attributes from high-resolution satellite data improve wood volume estimation in a semi-arid mountainous region?, Remote Sens. 8(7):540. https://doi.org/10.3390/rs8070540

Slocombe DS (1992) Environmental monitoring for protected areas: review and prospect. Environ Monit Assess 21(1):49–78. https://doi.org/10.1007/BF00400057

van der Werff H, van der Meer F (2016) Sentinel-2A MSI and Landsat 8 OLI provide data continuity for geological remote sensing. Remote Sens 8(11). https://doi.org/10.3390/rs8110883

Vargas L, Remme RP, Vargas L, Hein L, Remme RP (2017) Accounting for ecosystem assets using remote sensing in the Colombian Orinoco River Basin lowlands the Colombian Orinoco River Basin lowlands. J Appl Remote Sens 11(2). https://doi.org/10.1117/1.JRS.11.026008

Vincini M, Calegari F, Casa R (2016) Sensitivity of leaf chlorophyll empirical estimators obtained at Sentinel-2 spectral resolution for different canopy structures. Precis Agric 17(3):313–331

Xu C, Li Y, Hu J, Yang X, Sheng S, Liu M (2012) Evaluating the difference between the normalized difference vegetation index and net primary productivity as the indicators of vegetation vigor assessment at landscape scale. Environ Monit Assess 184(3):1275–1286. https://doi.org/10.1007/s10661-011-2039-1

Electronic Sources

AOPK: Regionální pracoviště Jižní Čechy. Agency of Nature Conservation. Available at: http://trebonsko.ochranaprirody.cz/, last access: 30.5.2019

CBD (2003) Methods for evaluating ecosystem integrity and monitoring ecosystem response. Available at: https://www.cbd.int/doc/pa/tools/Methods%20for%20Evaluating%20Ecosystem%20Integrity%20and%20Monitoring%20Ecosystem%20Respons.pdf, last access: 30.5.2019

Convention on Biological Diversity Aichi targets: Avalilable at: https://www.cbd.int/sp/targets/, last access: 30.5.2019

ESA SNAP: SNAP – Science Toolbox Exploitation Platform. European Space Agency. Available for download at: http://step.esa.int/main/download/, last access: 30.5.2019

ESA: Scihub Copernicus geoportal. European Space Agency. https://www.scihub.copernicus.eu, last access: 30.5.2019

USGS: Earth explorer geoportal, United States Geological Survey. https://earthexplorer.usgs.gov/, last access: 30.5.2019
Weiss, M., and Baret, F.: S2ToolBox Level 2 products: LAI, FAPAR, FCOVER, 53, 2016. Available at: http://step.esa.int/docs/extra/ATBD_S2ToolBox_L2B_V1.1.pdf, last access: 30.5.2019

Social and Economic Conditions for Wilderness Protection in Europe: Case of Šumava National Park, Czech Republic

Eva Cudlínová, Miloslav Lapka, and Martin Šlachta

Abstract Wilderness protection in Sumava National Park (NP) is full of conflicts. There are conflicts regarding the non-intervention zones, conflicts in relation to periodically recurring bark beetle calamities and conflicts with a new zonation proposal suggesting to increase the area of non-intervention zones up to 27.7% of the total area of NP. In general, all these conflicts described above are based on the cultural construction of wilderness connected with the historical picture of Sumava from the beginning of the nineteenth century. There occur some key duality points that exist in endless debates about wilderness in Sumava NP. showing the power of this cultural construction and its misuses. These dualities could be labeled as wilderness under human control. Cultural construction of wilderness in Sumava NP. is static and it is based on consumerism. A long-term perspective is often missing in the debate about the future of the Sumava wilderness. Patience with natural long-term processes is missing too. We suppose that case of Sumava National Park could serve as an example how important is the knowledge of historical, cultural and social-economic context of wilderness, for solving of existing difficulties and conflicts with wilderness protection.

1 Introduction

Nowadays we are used to speaking about the protection of nature, environmental protection or even wilderness protection as something inseparable from our life. The protection of nature is even a part of the 17th Sustainable Development Goal (United Nations 2019).

There are three fatal questions—"what, why and how to protect?" The first two questions "what kind of nature to protect and what are the reasons for protecting" are dealing with societal values and our relationship with nature. The last question

E. Cudlínová (✉) · M. Lapka · M. Šlachta
Department of Regional Management, Faculty of Economics, University of South Bohemia in České Budějovice, České Budějovice, Czech Republic
e-mail: evacu@ef.jcu.cz; mlapka@ef.jcu.cz; mslachta@ef.jcu.cz

L. Westra et al. (eds.), *Ecological Integrity in Science and Law*,
https://doi.org/10.1007/978-3-030-46259-8_15

"how to manage the protection of nature" depends on the legal and economic conditions of society.

In our chapter we would like to show how the three questions mentioned above are current in terms of the practical ongoing conflicts regarding the protection of Šumava National Park in the Czech Republic, as well as in the cultural construction of wilderness.

2 Defining of Wilderness in the USA and Europe

2.1 Wilderness in the US

Historically, the word was derived from the notion of *wildness* or that, which is not uncontrollable by humans. The word's etymology is from the Old English *wildeornes,* which derives from *wildeor* wild beast (from wild + *deor* beast, DEER) (The Collins English Dictionary 2000). During the European settlement of America the wilderness was seen as something to be feared as illustrated by one settler in the early 1600s, who stated that, "*Wilderness is a dark and dismal place where all manner of wild beasts dash about uncooked.*" Indeed, early settlers struggled to clear and cultivate natural lands as a way of civilizing wild America (Worboys et al. 2015).

According to Nash's famous book "Wilderness and the American Mind" (Nash 1982), "Wilderness" is a cultural human construction. The wilderness concept only exists in the minds of humans. Whilst this may be so, ironically, in order for there to be wilderness, there needs to be some humans in non-wilderness places defining what is wilderness and what is not (Neill 2003). Nash's book "Wilderness and the American Mind" has been rated in the top 100 most influential books in America (the first edition 1967 and the fifth edition 2014).

The United States was the first country in the world to define and designate wilderness areas through law, namely, the 1964 Wilderness Act. In the US Wilderness Act, wilderness is defined as "an area where the earth and its community of life are untrammelled by man, where man himself is a visitor who does not remain" (Wilderness Connect 2019).

Despite of fact of it being more of a metaphorical than scientific definition, we can see a dualism between man and nature, between culture and nature which is typical of the Western anthropocentric world-view. From this point of view, the place where we are is the place where nature is not (Nash 1982; Oelschlaeger 1991).

Although the Wilderness Act is a single piece of guiding legislation for all wilderness areas, many other laws affect the wilderness. Some laws have designated additional wilderness areas, subsequent to the initial 54 areas designated under the Wilderness Act. Others guide management, such as the Clean Air Act and the National Environmental Policy Act (Dawson and Hendee 2009).

2.2 *Wilderness in Europe*

A political milestone in the EU's definition of wilderness was the adoption of the "European Parliament Resolution on Wilderness in Europe" in February 2009.

The whole process of creating a definition of wilderness in Europe lasted nearly four years, from 2009 until 2013. A definition of wilderness had been formulated for the Conference, in Prague in May 2009, and the Wilderness Working Group (WWG) of the Wild Europe Initiative was established to develop this. The consensus definition of wilderness in a European context reads as follows:

> A wilderness is an area governed by natural processes. It is composed of native habitats and species, and large enough for the effective ecological functioning of natural processes. It is unmodified or only slightly modified and without intrusive or extractive human activity, settlements, infrastructure or visual disturbance (Wild Europe 2019).

2.3 *Comparison of Wilderness Protection in the US and Europe*

As already mentioned above, there are different definitions of wilderness in the US and Europe, and furthermore, differences exist in the definition and management of national parks.

The first and most famous national park worldwide is Yellowstone National Park, which was established in March 1872 and is rightly considered a turning point in territorial conservation across the world. It was proclaimed not for the protection of flora and fauna, but for unique geomorphological phenomena, especially geysers, and for the enjoyment of Americans who could use it for recreation.

A similar model of a national park as being a large area without zoning, with almost unlimited access for tourists can be found in many locations around the world, especially in Africa, Latin America and Australia (Sterling 2002). The first national parks were also established in Europe in places with a relatively low population density in Sweden in 1909 (EEA 2012). However, most national parks in Europe represent smaller areas with some cases of human management. In such a situation, zoning is very important for the protection of the living components of the ecosystems within the NP.

The definition of wilderness and the national or regional conditions as regards history, culture, economic and social conditions differ from state to state and region to region. We can observe how these conditions influence the protection of wilderness as in the case of the Šumava National Park in Czech Republic.

3 Wilderness Protection in Šumava National Park

Šumava Mts. as a region has a specific history concluded by the proclamation as a National Park in 1991. This territory was historically settled by German populations after the colonizers were invited by the Czech nobility in the Middle Ages. The situation changed after the Second World War, when the German population was resettled and new inhabitants from Romania and other countries of Eastern Europe came to live there for a short time. Until the fifties of the twentieth century this area served as a military zone in the border area between East and West Europe. Large areas along the state border to Germany and Austria were of limited or fully restricted access up to 1989 and a semi-natural landscape called the Green Belt gradually developed there. Due to this fact, 40 years without economic activity, this region conserved relatively pristine nature and could be proclaimed as a NP in 1991.

The main purpose of NP establishment was to protect a mosaic of unique peat bogs, spruce and beech forests, mountain meadows, wild rivers and glacial lakes. As one of the largest continuous forest complexes in Central Europe, it is called the Green Heart or Green Roof of Europe. Most of this area is covered by old spruce monocultures that in the eighteenth century replaced the former forest used as wood for the glass industry (Dudák et al. 2003).

Protective conditions of the territory were solved by the zoning of NP territory in three zones with different degrees of protection and management. Zone I (13% of the NP area) is strictly natural and includes the most valuable and stable areas with natural ecosystems like primeval forest remnants, wetlands and bogs. The territory of zone I is intended to be left to natural development without human influence.

Zone II (82% of the NP area) comprises the remaining majority of forest and other ecosystems, which were altered and heavily damaged and genetically inappropriate with varying degrees of composition and state as compared to the original ones. The aim of all activities is to maintain the natural balance and gradually bring the existing ecosystems closer to their natural state.

Zone III (5%) includes areas significantly modified by human activities and concentrated buildings. The aim is to maintain and promote the use of this zone for permanent housing, services, agriculture, tourism and recreation. This concept of protection and non-intervention (Zone I) is contained in the legislative Acts establishing the NP (National Park Šumava 2019).

3.1 Ongoing Main Conflicts

3.1.1 The Non-Intervention Zones

The non-intervention zones have entered into public debate and have been politicized in relation to periodically recurring bark beetle calamities. Conservationists, scientists and the administration of NP Šumava prefer to leave nature to itself in the strictly natural zones areas in order to cope with the calamity so as to allow its natural regeneration and reforestation. Foresters and municipalities, on the other hand, prefer intervention, i.e. cutting down and then planting new trees. They believe that Šumava has lost its status as wilderness due to the precedent of human interventions and cannot cope by itself with the current situation (Bláha and Kotecký 2015).

3.1.2 Bark Beetle Attack

The traditional forestry approach to reduce the activity of the bark beetle was applied in NP between 1995 and 2006. This involved the cutting down of infested trees even in non-intervention zones, which lead to forest thinning or creating clearings. In 1999, Friends of Earth, along with scientists and several local people, organized a peaceful blockade impeding logging in Trojmezná Old Growth Forest—the largest and most valuable remainder of an old growth montane spruce forest (Solar and Galland 2002). Because of the blockade of logging, Friends of Earth stopped extensive cutting in waterlogged natural forests around the Vltava river headspring, one of most attractive places for tourists. The NP director was removed by his superiors and felling in non-intervention zones was restricted in the following years. This approach was also applied in the case of windfalls which occurred after the extremely strong Kyrill storm in 2007 (Bláha and Kotecký 2015).

3.1.3 Zonation

A new zonation proposal has recently been prepared and is undergoing debate. According to the proposal, the area of non-intervention zones would rise to 27.7% of the total area of the NP. It would comprise the stands left without human intervention for more than 5 years, including areas disturbed by the storm Kyrill in 2007 and the subsequent bark beetle calamity (2009–2012). In view of a new enormous bark beetle calamity in other parts of Czech Republic as well as in other Central-European states, the administration of South Bohemia launched a campaign against the non-intervention management in Šumava NP once again in order to save the forests from "apocalypse" due to the spreading of bark beetles from non-intervention zones into neighbouring forests.

4 Discussion

Understanding and defining wilderness is not something stable like physical laws as it is changing over time and from nation to nation as was mentioned in the first part of the chapter. Wilderness means different things to different generations. The sixteenth century first settlers in the US perceived wilderness as something negative, whereas the present generation values and protects the same places. The shift in the cultural construction of wilderness is clearly visible in our case of the Šumava Mts., and ongoing conflicts of its management as well.

Adalbert Stifter in the novel Der Hochwald (1841), Hvozd in Czech/Deep Forest (Stifter 2003), Karel Klostermann in the novel V ráji šumavském (1893)/In Šumava Mts. Paradise (Klostermann 1999), describe places in the deep forest called "slatě" in Czech, or bogs. These are swamplands at high altitude in the Šumava Mts., and are nowadays the core zone of the National Park, strictly protected and accessible only by using a raised wooden walkway without touching any of the bog moss and other vegetation in order to protect the rich biodiversity. Bogs represent islands of life diversity in natural spruce deep forests for the management of the National Park. For the novelists mentioned above, bogs are places described like "desert", "death", "the dark", "lifeless place", and are full of awe and fear.

The dualism of culture and nature remains throughout time. It is visible on the monumental work of Czech painter, novelist, graphical designer and mystical artist Josef Váchal (1884–1969), who was fascinated by the wilderness of the Šumava Mts. In his monumental original book (20 kg weight!) from 1931, Šumava dying and romantic (Váchal 2008, in Czech) there is a romantic concept of the sublime as well as the concept of challenges to strong individualism, known from the US historical frontiers. These great artists and many others built the cultural construction of Šumava Mts. wilderness as something similar to other wilderness cultural constructions in the world.

The power of social constructivism is that externalization and objectivation are moments in a continuing dialectical process, confirmed by everyday behaviour, language and institutions. Those externalities such as the cultural construct of Šumava Mts. become internalities in terms of the sharing of the cultural consciousness "what does the wilderness of Šumava Mts. mean?" In the social world, "society is a human product. Society is an objective reality. Man is a social product" (Berger and Luckmann 1966). In these three sentences is the core of understanding the social construction of our human world, the cultural construction of wilderness included. There is nothing objective in terms of a scientific approach to wilderness and in this term the critics are right (Crist 2004). On the other hand, this construction is of an objective cultural environment and sharing experiences with wilderness, which cannot be omitted.

The new lesson from the Šumava Mts. cultural construction of wilderness consists in the way in which the construct could be misused. There are the strong individual interests of politicians, developers and local municipalities to profit from the externalities connected with the Šumava wilderness, disregarding any barriers, even in spite of the law of protecting national parks. This selfish battle for individual economic profit uses the cultural construct of the Šumava Mts. wilderness in many ways. There is the underlying fact that the wilderness construct is based on a more or less public sharing of a picture of the Šumava Mts. as a deep forest, with majestic spruces, green open pastures, small villages located in harmony with nature and people with experience and wisdom of heroic frontiers. The historical and social reality is another construct, apart from the environmental reality of bark beetle which rapidly dry and destroy non-indigenous even-aged stands.

This environmental fact opens the door for an attack against National Park management and argues for the free intervention in core zones on behalf of protection of the Green Šumava wilderness construction. To save Šumava Mts. means to maintain the whole area by human intervention and by the way to ensure individual economic and political benefits to a group of interested politicians and/or landowners.

There occurs following key duality points that exist in excited debate about wilderness in Šumava Mts. showing the power of this cultural construction and its misuses.

4.1 Wilderness Under Human Control

The paradox of dualism grows into a curious situation when many people and foresters do not believe in the power of the wilderness to regenerate itself without the help of man. This "help" means to cut down trees after a bark beetle attack, applications of pesticides, etc.

Dualism in the cultural construction of wildness in Šumava Mts. takes the paradoxical form of arguments against the existence of National Park as such, against the Park management as well as against the Ministry of Environment. It has gained political support in the post-communist economic lobby, including some majors of local villages in Šumava Mts. Nature needs our help fits into our cultural construction of what should be the "right" wilderness in Šumava Mts. Nature without human intervention is weak and nature in great transformation is not what we want to see and consume.

4.2 Wilderness Is Endangered by Strong Protection

The discourse in newspapers use wording like "bark beetle apocalypse" in headlines. This has been repeated several times by the governor of the South Bohemia region where the Šumava Mts. are located. "Thousands of dead hectares", "We can't handle calamity" are the other headlines in Czech newspaper (MF Dnes 2019a, b). The fact that the forest is spontaneously regenerated does not fit into this picture and even fake photographs of the core zones have been used.

4.3 Cultural Construction of Wilderness in Šumava Mts. Is Static

There are no natural processes like an extinction of one ecosystem, its transformation and natural restoration in the politically presented construction of Šumava Mts. In this construction, there are no natural disturbances such as destroyed forests after storms, bark beetle, fires or droughts. If it happens, it is called a catastrophe, crisis or apocalypse as mentions above. Wilderness without change does not exist, of course, but our romantic picture of how Šumava should look like does not allow for change.

4.4 Cultural Construction of Wilderness in Šumava Mts. Is Based on Consumerism

This point fits the philosophy of the modern economic lifestyle. Modern man needs to consume wilderness just now, with little regard for future generations. There is no time to wait hundred years for the regeneration of the forest, there is no patience with slow evolutionary natural processes. Arguments for human interventions in wilderness are based on our impatience.

All of these misuses are naturally interconnected and build a complicated picture of wilderness cultural construction in modern society. Despite the underlying fact of individual economic and political profit for many stakeholders involved in Šumava Mts., we are facing reality, which is objective in terms of real power of the wilderness construction.

5 Conclusions

Is there any solution? For an answer, we need to stress a geographical fact; we have no more space for some other wilderness in some other Šumava Mts. in central Europe. Every region is full of competing interests and the Šumava Mts. are now at the forefront of these competing conflicts.

A long-term perspective is often missing in the debate about the future of the Šumava wilderness. Patience with natural long-term processes is missing. On the other hand, it is necessary to ask about the historical, cultural and social-economic context of wilderness. However, if we are convinced we need wilderness, it means wilderness needs some space for its own development and narratives in our landscape as well in our cultural and economic values. It means to admit that wilderness has a right to be independent of human intervention but not isolated from human life. Bottom-up as well as top-down management of wilderness is not sufficient in the case of the Šumava Mts. It requires more multilevel governance, based on the modern cultural construction of wilderness and its values for modern man. The cultural construction of wilderness is not only an academic topic as we see in the case of conflicts and debates surrounding Šumava NP. The real power of culture construction and its misuse in public debate affects the future existence of this National Park.

Acknowledgments We would like to thank Schaefer Justin Calvin, BSc. for his help with the English review. The chapter was supported by the projects NAZV QK 1920391 Diversification of the Bioeconomy on Strategic Documents of the Forestry—Wood Sector as Basis for State Administration and the Design of Strategic Goals by 2030, Ministry of Agriculture of the Czech Republic and by the Czech Science Foundation (Grant no. 19-10694S).

References

Berger PL, Luckmann T (1966) The social construction of reality: a treatise in the sociology of knowledge. Anchor Books, Garden City

Bláha J, Kotecký V (2015) 'Spruce bark beetle in Šumava NP: a precedent case of EU Wilderness Protection, the role of NGOs and the public in wilderness protection' In: Watson A, Carver S, Krenová Z, McBride B (eds) Science and stewardship to protect and sustain wilderness values: Tenth World Wilderness Congress symposium; 2013, 4-10 October, Salamanca, Spain. Proceedings RMRS-P-74, Department of Agriculture, Forest Service, Rocky Mountain Research Station, Fort Collins

Crist E (2004) Against the social construction of nature and wilderness. Environ Ethics 26(1):5–24

Dawson CP, Hendee JC (2009) Wilderness management: stewardship and protection of resources and values. Fulcrum Publishing, Golden

Dudák V et al (2003) Šumava – příroda, historie, život/Šumava Mts.- nature, history, life. Baset, Prague, Czech Republic

EEA (2012) Protected areas in Europe – an overview. Publication Office of the European Union, Luxembourg

Klostermann K (1999) V ráji šumavském/In Šumava Mts. Paradise (orig. 1893), Nakladatelství lidové noviny, Prague, Czech Republic

MF Dnes (2019a) Kůrovcová apokalypsa/Bark beetle apocalypse. MF Dnes (September 7, 2019)

MF Dnes (2019b) Kraj chce sázet stormy/The Region wants to plant trees. MF Dnes (September 10, 2019)

Nash R (1982) Wilderness and the American mind. Yale University Press, New Haven

National Park Šumava (2019) Official website of the National Park Šumava. www.npsumava.cz/cz/. Accessed 21 Oct 2019

Neill J (2003) What is wilderness? Definition & discussion of wilderness as a human construction. www.wilderdom.com/wilderness/WildernessDefinition.html. Accessed 18 October 2019

Oelschlaeger M (1991) The idea of wilderness: from prehistory to the age of ecology. Yale University Press, New Haven

Solar M, Galland P (2002) Zpráva o misi IUCN / WCPA do Národního parku Šumava/Report on the IUCN/WCPA mission in the Šumava National Park. www.sumavanet.cz/www/data/com/iucn/iucn02.htm. Accessed 21 Oct 2019

Sterling EJ (2002) Conservation: definition and history. In: Eldridge N (ed) Life on earth: an encyclopaedia of biodiversity, ecology, and evolution. ABC-CLIO, Santa Barbara

Stifter A (2003) Hvozd/Der Hochwald/Deep Forest (orig. 1841), Vitalis, Prague, Czech Republic

The Collins English Dictionary (2000) Harper Collins, London

United Nations (2019) Sustainable Development Goals. https://www.un.org/sustainabledevelopment/. Accessed 21 Oct 2019

Váchal J (2008) Šumava umírající a romantická/Šumava dying and romantic (orig. 1931), Paseka, Prague, Czech Republic

Wild Europe (2019) Wild Europe definition of wilderness: The Importance of Practical Definitions. https://www.wildeurope.org/definitions/. Accessed 21 Oct 2019

Wilderness Connect (2019) The Wilderness Act, Public Law 88-577 (16 U.S.C. 1131-1136) 88th Congress, Second Session September 3, 1964. wilderness.net/learn-about-wilderness/key-laws/wilderness-act/default.php. Accessed 18 Oct 2019

Worboys GL, Lockwood M, Kothari A, Feary S, Pulsford I (eds) (2015) Protected area governance and management. ANU Press, Canberra

Is it Possible to Maintain a Biodiversity of Natural Habitats Under Global Change and Increasing Landscape Exploitation?

Pavel Cudlín, Vilém Pechanec, Marcela Prokopová, Lenka Štěrbová, Jan Purkyt, and Ondřej Cudlín

Abstract In a period of global change, we consider biodiversity as a feature that is crucial for nature resistance and resilience, which are prerequisites of an ecosystem's ability to mitigate the negative impacts of global change on human populations. However, biodiversity is also highly jeopardized by the climate and land use changes. Its decline is alarming and further deterioration is expected.

The goal of this chapter was to detect these changes in the Czech Republic. An analysis of Corine land cover data time sequences was performed to quantify the changes at the landscape level, while data of the Habitat layer of the Czech Republic from two time periods based on Natura 2000 mapping were compared to define changes in natural habitats. The Habitat Valuation Method was used to quantify the loss of biodiversity value on both levels.

The results of land cover development showed increased biodiversity value for the whole Czech Republic, mainly due to significantly increased areas of meadows and pastures at the expense of arable land and a smaller increase of forest area on abandoned soil. On the other hand, the biodiversity values of natural and near natural habitats decreased in favour of ruderalized habitats, mostly due to increasing pressure from landscape exploitation and climate change.

P. Cudlín (✉) · M. Prokopová · L. Štěrbová · O. Cudlín
Global Change Research Institute, Academy of Sciences of the Czech Republic, České Budějovice, Czech Republic
e-mail: cudlin.p@czechglobe.cz

V. Pechanec
Department of Geoinformatics, Life Science Faculty, Palacky University in Olomouc, Olomouc, Czech Republic

J. Purkyt
Global Change Research Institute, Academy of Sciences of the Czech Republic, České Budějovice, Czech Republic

South Bohemian University, České Budějovice, Czech Republic

L. Westra et al. (eds.), *Ecological Integrity in Science and Law*,
https://doi.org/10.1007/978-3-030-46259-8_16

1 Introduction

Nowadays the scientific community is increasingly focused on biodiversity and its development under global change conditions. Climate, land-use and biodiversity are all negatively impacted by the human population growth and activities. Climate and land use change negatively impact biodiversity not only directly, but also through the combinations they produce and whose effects are not yet sufficiently known (Scherber 2015). In addition, biodiversity is becoming increasingly important to us in the context of coming changes, as it is a prerequisite for the adaptation to and mitigation of the negative effects of global change.

1.1 How Biodiversity Mitigates the Impact of Global Change?

Biodiversity has some capacity to stabilize a landscape exposed to climate and land use changes. There is a predominant view among ecological scientists that resilience is at several scales, related to some aspects of diversity, mainly through redundancies, response diversity and spatiality (Adger 2006; Folke 2006). Lavorel et al. (2015) list vegetation structural diversity and response diversity among the main mechanisms underpinning ecosystem services and reassembly of ecological communities, while Huyghe et al. (2012) emphasize functional diversity.

At individual levels, the effect of genetic diversity promotes the adaptability of wild species (Reusch et al. 2005) as well as agricultural species (Wu and Wu 2013). On the community level, species diversity is related to higher redundancy, which brings differing impacts and responses to environmental change or perturbation (Duffy 2002; Swift and Hannon 2010). This variety is crucial for response diversity; it determines the resilience, adaptability and flexibility of ecosystems (Duffy 2002; Walker and Salt 2006) and underpins the reassembly of communities (Lavorel et al. 2015). Biodiversity also decreases the risk of pathogens dispersal (Chapin 3rd et al. 2000; Hooper et al. 2005) as well as invasive species (Elton 1958 in Chapin 3rd et al. 2000), which are expected to increase with global warming. Thus, biodiversity loss would make the consequences of global changes more fatal.

1.2 Land Use and Land Cover Change As a Threat to Biodiversity

Land use (LU) and land cover (LC) change is one of the most serious causes of habitat loss and consequently biodiversity decline (Sala et al. 2000; Falcucci et al. 2007). Both theoretical and field studies suggest that many species are not able to cope with such changes without conservation measures (e.g. Thuiller 2007). There are various ways in which land use change affects biodiversity. The most visible is an expansion of agricultural or urban land into natural habitats (Frondoni et al. 2011; Smiraglia et al. 2016) causing a loss of entire habitats.

However, agricultural intensification and urbanization have also indirect effects, such as fertilization, pesticide pollution or N deposition that cause degradation of habitats with a subsequent decrease in biodiversity (Erisman et al. 2015; Simkin et al. 2016). Another negative effect of land use change is fragmentation that affects the biodiversity of the remaining natural or semi-natural habitats by causing a decline in species abundance and richness (Van Eetvelde and Antrop 2009; Oliver and Morecroft 2014) or changes in water regime (mainly by urbanization and agriculture intensification), which can lead to water stress or floods (Kedziora 2010).

Abandonment of land and the lack of management resulting in shrub and tree encroachment into old pastures and cultivated lands can (especially in Central Europe) result in either increased or decreased biodiversity; the former mostly in case of abandoned arable land (Rehounkova and Prach 2007), the later predominantly in the case of abandoned near natural grasslands (Plieninger 2006; Honrado et al. 2017).

1.3 Climate Change As a Threat to Biodiversity

Climate change poses a threat to biodiversity by affecting mean climate parameters and the frequency and intensity of climate extremes (Jentsch et al. 2007). Some species will face the risk of extinction, while others will survive through combinations of shifting their distribution to track the climate, persisting in climatic refugia or adapting to the new conditions (Dawson et al. 2011).

Changes in climate can also trigger changes in biodiversity by creating opportunities for previously innocuous alien species (Thuiller 2007). It has also been commonly observed that climate change will favour forest pest species, since the survival of many arthropods depends on low temperature thresholds (Williams and Liebhold 1995). Fungi or pathogens also benefit from changed climatic conditions due to increased temperatures and higher water stress of plants (Jactel et al. 2012). Climatic conditions of high temperatures and low humidity, together with the possibly decreased vitality of vegetation, can increase the risk of fire (Mouillot et al. 2002).

1.4 Multiple Effect of Climate and Land Use Changes to Biodiversity Loss

It has been proved that land cover affects the climate (1) through carbon sequestration, (2) thanks to vegetation characteristics such as albedo, roughness length or leaf area (Copeland et al. 1996). For the Czech Republic, the prediction of carbon storage under climate change was carried out, based on future land use changes and evaluation of changes in human impacts on biomass production and carbon cycle (Pechanec et al. 2018). In addition to the interaction of climate and land use, we can also expect a multiplication of negative impacts on the biodiversity when combining these drivers. A combination of climate change, species invasions and reduced areas of natural habitat is likely to promote biotic homogenization in biodiversity hotspots in particular and foster unpredictable interactions between plants, animals and microorganisms (Thuiller 2007).

Population responses to extreme climate events, such as drought, are likely to be affected by habitat quality, area, configuration and heterogeneity (Fischer and Lindenmayer 2006; Oliver and Morecroft 2014). The most discussed is the impact of land use on a species ability to shift their distribution in response to climate warming (Hill et al. 2001; Travis 2003). An intensively managed landscape may restrict species migration and the ability to cope with climate change by tracking of suitable conditions (Thomas et al. 2004). Also, a combination of low habitat heterogeneity and climate change can negatively impact biodiversity, because greater habitat and topographic heterogeneity provide broader microclimatic gradients necessary for species survival (Oliver and Morecroft 2014).

Recent studies also stress the important influence of land-cover change in accelerating invasions (Polce et al. 2011). Open or disturbed areas related to land use changes are often favoured by non-native species adapted to take advantage of resource release (González-Moreno et al. 2014). Moreover, certain land-use changes increase the fragmentation and isolation of forest landscapes, which are more invaded than large continuous forests (Malavasi et al. 2014).

1.5 The Spatial Data, Persistence of Land Cover/Habitats

LU changes are often studied by using temporal satellite images. A spatial data time series enables us to compare LC or habitat composition in individual years. Using the proportion of natural/near natural habitats, we can estimate the relative biodiversity in individual years, because natural parts in the landscape increase its ecological stability through a higher amount of resources, species richness, and population sizes (Hodgson et al. 2011). Trajectory construction is more suitable for tracking changes (Frondoni et al. 2011) while persistence (continuity) analysis can indicate stability, integrity or habitat quality (Skaloš et al. 2015; Pitkanen et al. 2016). Conversely, the relative amount of transition land can be used as an indicator of long-term changes in the landscape (Wästfelt and Arnbergb 2013). Land use history and habitat continuity play crucial roles in determining the degree of diversity (Lovett et al. 2005), as been confirmed in grasslands (Cousins et al. 2009) and forests (Nordén et al. 2014). The number and location of stable (unchanged) polygons with a long continuity of development can detect areas with higher ecological quality and consequently higher biodiversity (Käyhkö and Skånes 2006). Skaloš et al. (2015) used unchanged forest polygons to assess the quality of forest integrity, while Pitkanen et al. (2016) used stable polygons for a stability and human pressure assessment.

2 Methods

2.1 *Spatial Analysis of Land Use and Biodiversity Changes*

The spatial data available to assess land use and biodiversity change at the landscape level were based on CORINE Land Cover (CLC) satellite data and on data from the field Habitat Mapping of the Natural Conservation Agency of the Czech Republic (NCA CR). The CLC includes 29 land cover (LC) classes and has been done for 1990, 2000, 2006, 2012, and 2018. LC analyses were carried out for the whole monitored period (1990–2018). This consisted of a comparison of the area of CLC categories in individual years, construction of development trajectories and selection of persistent areas.

Natural and near natural habitats were analysed using data from the Habitat mapping of NCA CR. Data from two time periods were available, the original mapping from 2004 and the revision from 2014. The areas of individual habitats were compared between 2004 and 2014 and the persistence proportion of each habitat was also identified. Subsequently, the changes were expressed in point and monetary values using the Habitat valuation method (HVM) (Seják et al. 2003; Cudlín et al. 2005).

2.2 *Biodiversity Valuation Using the Habitat Valuation Method*

We valuated habitats as biodiversity carriers by HVM. This method was based on the original Hessen method used Germany to evaluate ecosystem loss. The first version of the method valuated 192 habitat types in the Czech Republic: 139 natural and near natural habitats from the Habitat Mapping of NCA CR and 53 more unnatural habitats (Seják et al. 2003; Seják and Cudlín 2010). During 2016–2018, the method was further developed so that categories of non-natural habitats are based on vegetation units rather than land use and the number of unnatural habitats was reduced to 38 habitats. The relative assessment of habitats, expressed as a point value per 1 m^2, was based on eight criteria: habitat maturity, habitat naturalness, diversity of layers-structures, diversity of plant species, rareness of habitat, rareness of plant species, vulnerability of habitat type, endangerment of quantity, and quality of habitats. These criteria were assessed for every habitat type (139 natural and near natural and 38 unnatural habitats). Using the calculation formula, we obtained the relative value of each type of habitat (Seják et al. 2018). This value was derived from 182 restoration projects realized in the time period 2007–2016 and presents the cost that the government paid in the frame of restoration programs to increase the ecological quality of 1 m^2 of habitat by 1 point. The financial value of one point is about 1 EUR (Seják et al. 2018).

The HVM valuation method was applied to the CLC data sequence from 1990 to 2018 for the whole Czech Republic to obtain the development of habitat value during the selected period. The application of HVM to the Corine LC data was enabled by converting the CLC categories into habitat categories. This conversion was based on overlapping analyses which determined the average representation and representation of individual habitat types according to the HVM classification for each CLC category. For better clarity of the HVM results, the CLC categories were grouped into formations.

The HVM method was also applied to data from the Habitat mapping of NCA CR from 2004 and 2014, which included only natural and near natural habitats. Again, the habitat types were grouped into formations and the values from the two mapping times were compared.

3 Results

3.1 *Changes in the Area and Biodiversity Value of the Land Cover Categories in the Czech Republic Since 1990*

The most significant change in the last three decades concerning land cover is the conversion of arable land to pastures. This result was partly due to the end of the communistic period (1989), with the resulting end of centralized farming on agricultural land and privatization of previously collectivized land, but largely it was due to the introduction of state subsidies for this conversion in 2000 whose aim was to rectify intensive farming in the communist era and protect the soil against erosion. Between 2000 and 2006, the change of arable land to pastures was detected on 1869 km^2, while it was only 124 km^2 between 2012 and 2018 (Table 1).

Recently, urban area expansion has become dominant, especially the growth of the discontinuous urban fabric (increase of 368 km^2 since 1990) and industrial and commercial fabrics at the expense of arable land. The increase of all urban land since 1990 is almost 557 km^2. The highest increase was between 2000 and 2006, which later slowed down, but recently, since 2012, it has started growing steeply again (Fig. 1).

The amount of arable land decreased significantly, by 6836 km^2 since 1990, being mostly converted to the category "Pastures" (all cultural meadows and pastures). However, the role of urbanization in this process has been recently increasing (Table 1, Fig. 2).

The area of pastures and meadows increased from 2528 km^2 in 1990 to 8068 km^2 in 2018. However, the area of natural grasslands decreased from 405 km^2 in 1990 to 252 km^2 in 2018, which means that it represented (according to CLC data) only 3% of all grasslands in 2018, while in 1990 it was almost 14% (Table 1, Fig. 3).

According to the HVM based on the Corine LC data, the biodiversity habitat value of the Czech Republic (1990–2018) has increased since 1990, mainly due to the transfer of a large area of arable land to pastures and meadows after 2000. These meadows were mostly intensively managed and therefore their biodiversity was not

Table 1 Area of Corine LC categories in the Czech Republic between 1990 and 2018 (km^2)

	Area in years				
Land cover categories	1990	2000	2006	2012	2018
Continuous urban fabric	14.64	14.64	15.67	15.67	15.7
Industrial or commercial units	521.2	547.7	602.1	631	656.5
Road and rail networks	48.1	52.7	62.6	72.1	71.9
Discontinuous urban fabric	3579	3626	3784	3825	3947
Total urban areas	*6152.41*	*6240.95*	*6469.91*	*6556.03*	*6709.17*
Non-irrigated arable land	35.541	32.622	29.892	28.911	28.705
Complex cultivation patterns	415	430	476	473	474
Agricultural land with natural elements	6736	6748	7079	7114	7128
Total agricultural land	*42.693*	*39.799*	*37.447*	*36.578*	*36.307*
Pastures	2528	5317	7186	7944	8068
Natural grasslands	405	392	262	257	252
Total grasslands	*2932*	*5709*	*7448*	*8201*	*8320*
Broad-leaved forests	2495	2527	2783	2839	2834
Coniferous forests	16.552	16.993	17.227	17.126	16.658
Mixed forests	5855	6042	6174	6337	6437
Total forests	*24.902*	*26.184*	*26.184*	*26.302*	*25.929*

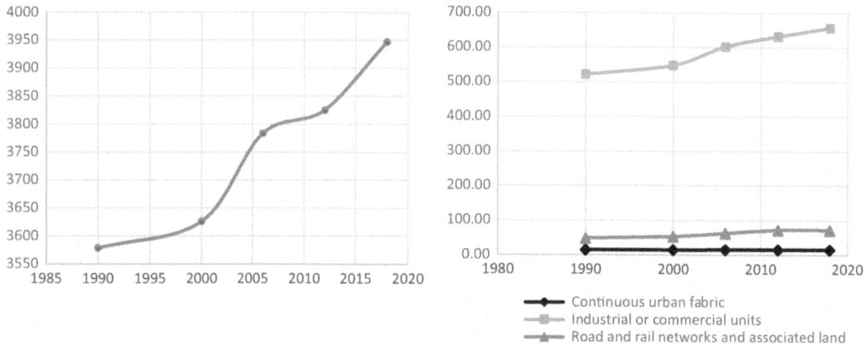

Fig. 1 The growth of discontinuous urban area (on the left side) and other types of urban area (on the right side)

high. However, thanks to the enormous area converted, this change increased the overall biodiversity value of the entire Czech Republic; the trend of biodiversity value development is very similar to the shape of grassland area development (Fig. 4).

The transfer of unmanaged land into forest and, thanks to restoration projects, the increase in the ratio of scattered trees, alleys and hedgerows in some areas, which led to part of arable land being transferred into the category Agricultural land with natural elements, also contributed to this positive trend.

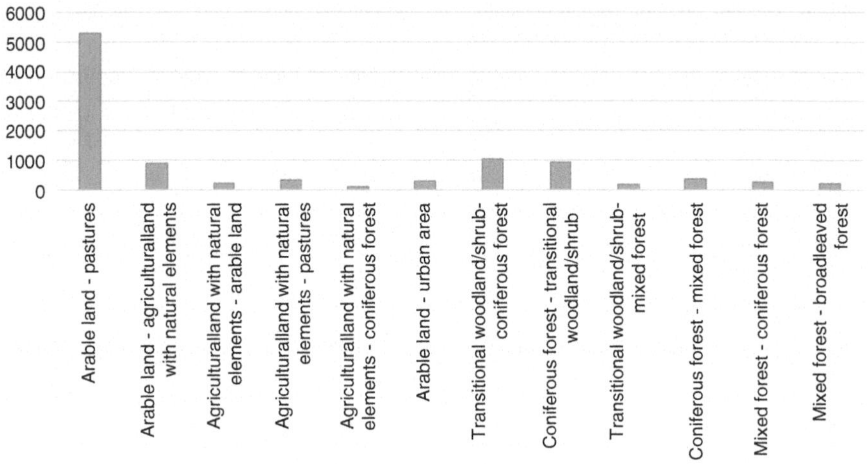

Fig. 2 The main changes of Corine LC categories from 1990 to 2018 (km²)

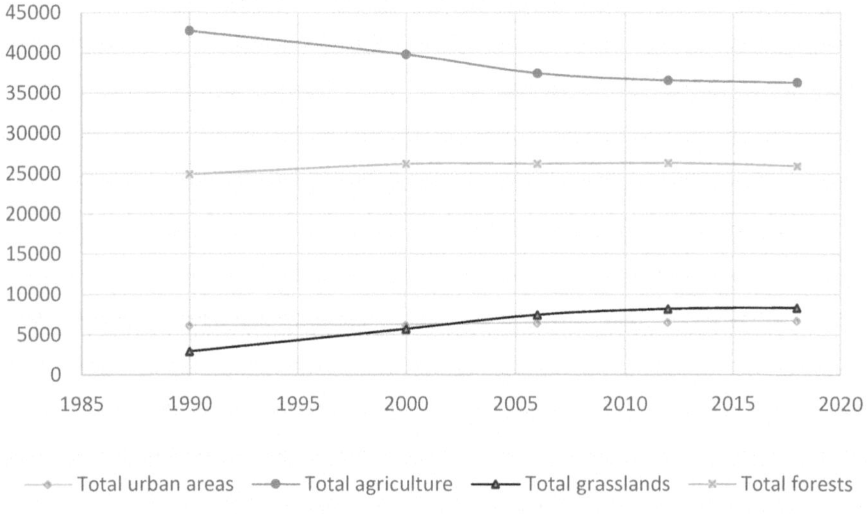

Fig. 3 Area of summarized Corine LC categories between 1990 and 2018 (km²)

3.2 Changes in the Area and Biodiversity Value of Natural and Near Natural Habitats Between 2004 and 2014

Unfortunately, the results are much less optimistic for natural and near natural habitats, as based on the data from the Habitat Mapping of NCA CR. Almost all natural habitat types decreased between 2004 and 2014. Natural and near natural grasslands

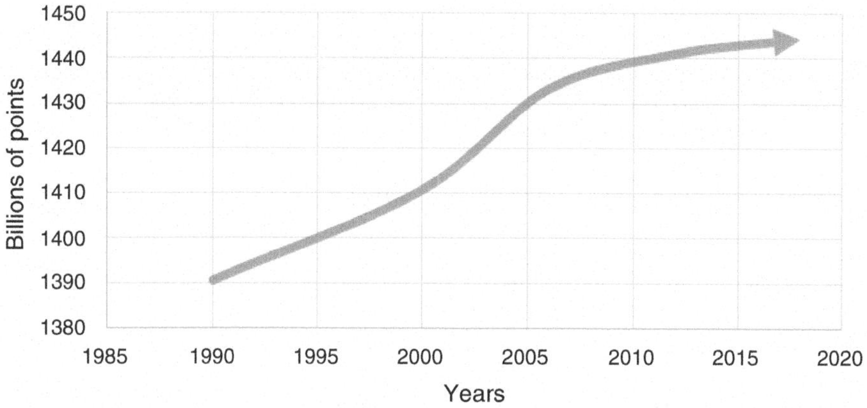

Fig. 4 Development of biodiversity point value of the Czech Republic counted by HVM (Habitat Valuation Method) based on Corine LC data

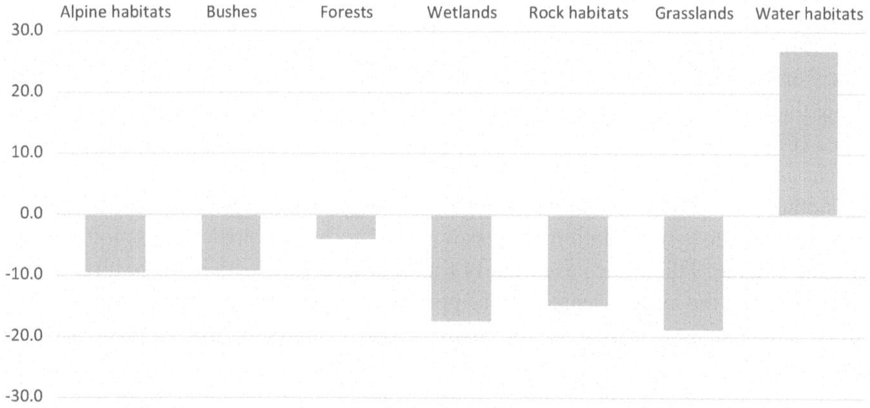

Fig. 5 Proportional decrease of natural and near natural habitats in individual formations between 2004 and 2014 (% from 2004)

showed the most serious decline (almost 337 km^2; 18.7%), followed by natural forests with a loss close to 158 km^2 (3.9%). Overall, the area of natural habitats decreased by 8.6%, which is a rather alarming result. The only exception are water habitats, which increased by almost 27%, probably due to subsidies for the restoration of wetlands, ponds and streams (Fig. 5).

This situation was reflected in the monetary valuation. HVM points of natural and near natural habitats decreased significantly between 2004 and 2014. In monetary terms, the decrease of total habitat value in the whole Czech Republic was calculated at 32,075 billion € (Table 2).

Table 2 Point values according to HVM for individual formations between 2004 and 2014

Formation	Point value in 2004 (billions of points)	Point value in 2014 (billions of points)	Increase (+)/ decrease (−) of point value (billions of points)	Increase (+)/decrease (−) of point value (% from point value in 2004)
Alpine	1.972	1.782	−0.189	−9.6
Bushes	7.442	6.694	−0.749	−10.1
Forests	216.132	204.658	−11.474	−5.3
Wetlands	9.397	7.835	−1.562	−16.6
Rock habitats	1.241	1.060	−0.181	−14.6
Grasslands	80.107	65.107	−15.000	−18.7
Water habitats	3.722	5.407	1.685	45.3
Total natural habitats	*320.013*	*292.544*	*−27.469*	*−8.584*

3.3 Persistence of Natural/Near Natural Habitats

In addition to the comparison of habitat areas in these 2 years of mapping, a persistence analysis was conducted to determine if habitats remained in place. The results show that, for most habitats, less than half of the original area in 2004 remained in place (Fig. 6). The high proportion of this phenomenon could be caused by natural succession. The least persistent are wetlands, because they represent a young successional state prone to natural change. However, a large proportion of habitats disappeared due to degradation processes caused by human pressure, exploitation or climate change. This is the case mainly for natural forests, with some natural beech forests having the largest degree of disappearance. Another reason is a lack of traditional management, which is the main cause of degradation of near natural grasslands.

4 Discussion

We obtained completely different trends using Corine LC data covering the whole Czech Republic, which represent changes in land cover, versus when focusing only on natural and near natural habitats as mapped by the Habitat Mapping of NCA CR (habitat representation reflecting biodiversity changes).

With the first, there is a positive trend caused by the large increase of areas with relatively low biodiversity quality, while with the other, there is a rapid decrease of natural and near natural habitats.

According to the Habitat Valuation Method, the bigger area of habitats with relatively low biodiversity quality can be equal to a smaller area of habitat with high quality. But is it correct to make such comparisons? A large area of average or even low biodiversity value can support ecosystem functions, but on the other hand, a

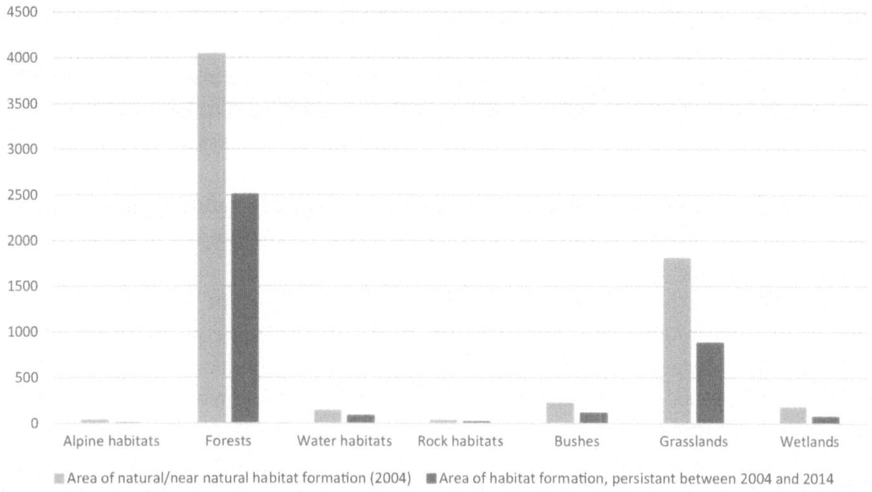

Fig. 6 Area of persistant habitats (km^2) in 2014 compared to their original size in 2004 (according to formations)

small area of natural habitats with the highest biodiversity and integrity represent the reservoir of species and their conservation should have the highest priority. As noted by Young (2000), "although restoration can enhance conservation efforts, restoration is always a poor second to the preservation of original habitats". So how to set our priorities? Walker and Salt (2006) suggest that, to support ecosystem resilience, we should promote diversity in all its dimensions, from biological to economic, and encourage multiple components and resource uses to balance and complement homogenizing trends. Heller et al. (2015) advise to create conservation networks that capture the full range of climatic diversity in a region in order to improve the resilience of biotic communities to climate change compared to networks that do not. Most probably, people will have to consider a new view of landscape protection and follow the New Wilderness Concept: 'Protected areas that are usually large, unmodified or slightly modified areas, retaining their natural character and influence, without permanent or significant human habitation, which are protected and managed so as to preserve their natural condition' (Dudley 2013, p. 14). The goal is to maintain the highest integrity of ecosystems, wildlife, and sacred and traditional cultural-use sites, to preserve intrinsic wilderness values, while sustaining human values (Casson et al. 2016).

5 Conclusions

Although the value of habitats calculated for the whole Czech Republic increased (above all due to the significant increase of meadows and pastures at the expense of arable land and some increase of forest area on abandoned soil), we must not be misled by this result. The overall biodiversity and species richness of the Czech

Republic is concentrated mainly in natural habitats and their area and value have been continuously decreasing in favour of ruderalized habitats, mostly due to increasing pressure from landscape exploitation and a strong climate change impact. The persistence of natural and near-natural habitats was found to be quite low.

It is not possible to stop biodiversity loss under the present socio-economic conditions. Therefore, nature protection has a very difficult position. Among many general reasons, two could be mentioned: (1) man is protecting nature against himself and (2) only the stronger can protect the weaker (the problem up to now persisting in some developing countries).

Application of the proposed IUCN "new wilderness concept" is one possibility how to mitigate biodiversity loss (Casson et al. 2016). This concept is based on delimitation of large-scale areas (if possible) that have remained relatively undisturbed or abandoned, where ecological processes can function without human intervention at least in the future.

References

Adger WN (2006) Vulnerability. Glob Environ Change 16:268–281

Casson SA, Martin VG, Watson A, Stringer A, Kormos CF, Locke H, Ghosh S, Carver S, McDonald T, Sloan SS, Merculieff I, Hendee J, Dawson C, Moore S, Newsome D, McCool S, Semler R, Martin S, Dvorak R, Armatas C, Swain R, Barr B, Krause D, Whittington-Evans N, Gilbert T, Hamilton L, Holtrop J, Tricker J, Landres P, Mejicano E, Gilbert T, Mackey B, Aykroyd T, Zimmerman B, Thomas J (2016) Wilderness protected areas: management guidelines for IUCN category 1b protected areas. IUCN, Gland, p 92

Chapin FS 3rd, Zavaleta ES, Eviner VT, Naylor RL, Vitousek PM, Reynolds HL, Hooper DU, Lavorel S, Sala OE, Hobbie SE, Mack MC, Díaz S (2000) Consequences of changing biodiversity. Nature 405:234–242

Copeland JH, Pielke RA, Kittel TGF (1996) Potential climatic impacts of vegetation change: a regional modeling study. J Geophys Res 101:7409–7418

Cousins SAO, Lindborg R, Mattsson S (2009) Land use history and site location are more important for grassland species richness than local soil properties. Int J Bot Mycol 27:483–489

Cudlín P, Prokopová M, Francírková T, Burešová R, Smrž T, Boucníková E (2005) System NATURA 2000 utilization for purposes of biotope valuation. Ekológia 24:52–68

Dawson TP, Jackson ST, House JI, Prentice IC, Mace GM (2011) Beyond predictions: biodiversity conservation in a changing climate. Science 332:53–58

Dudley N (ed) (2013) Guidelines for applying protected area management categories. IUCN, Gland

Duffy JE (2002) Biodiversity and ecosystem function: the consumer connection. Oikos 99:201–219

Erisman JW, Galloway JN, Dice NB, Sutton MA, Bleeker A, Grizzetti B, Leach AM, de Vries W (2015) Nitrogen: too much of a vital resource. Science Brief. WWF Netherlands, Zeist

Falcucci A, Maiorano L, Boitani L (2007) Changes in land-use/land-cover patterns in Italy and their implications for biodiversity conservation. Landscape Ecol 22:617–631. https://doi.org/10.1007/s10980-006-9056-4

Fischer J, Lindenmayer DB (2006) Beyond fragmentation: the continuum model for fauna research and conservation in human-modified landscapes. Oikos 112:473–480

Folke C (2006) Resilience: the emergence of a perspective for social-ecological systems analyses. Glob Environ Change 16:253–267

Frondoni R, Mollo B, Capotorti G (2011) A landscape analysis of land cover change in the municipality of Rome (Italy): spatio-temporal characteristics and ecological implications of land cover transition from 1954 to 2001. Landsc Urban Plan 100:117–128

González-Moreno P, Diez JM, Richardson DM, Montserrat V (2014) Beyond climate: disturbance niche shifts in invasive species. Glob Ecol Biogeogr 24(3):360–370. 1–11

Heller N, Kreitler J, Ackerly DD, Weiss SB, Recinos A, Branciforte R, Flint LE, Flint AL, Micheli E (2015) Targeting climate diversity in conservation planning to build resilience to climate change. Ecosphere 6:1–20

Hill J, Collingham Y, Thomas C, Blakeley D, Fox R, Moss D, Huntley B (2001) Impacts of landscape structure on butterfly range expansion. Ecol Lett 4:313–321

Hodgson JA, Moilanen A, Wintle BA, Thomas CHD (2011) Habitat area, quality and connectivity: striking the balance for efficient conservation. J Appl Ecol 48:148–152

Honrado JP, Lomba A, Alves P, Aguiar C, Monteiro-Henriques T, Cerqueira Y, Monteiro P, Barreto Caldas F (2017) Conservation management of EU priority habitats after collapse of traditional pastoralism: navigating socioecological transitions in mountain rangeland. Rural Sociol 82:101–128

Hooper DU, Chapin FS III, Ewel JJ, Hector A, Inchausti P, Lavorel S, Lawton JH, Lodge DM, Loreau M, Naeem S, Schmid B, Setälä H, Symstad AJ, Vandermeer J, Wardle DA (2005) Effects of biodiversity on ecosystem functioning: a consensus of current knowledge. Ecol Monogr 75:3–35

Huyghe C, Litrico I, Surault F (2012) Agronomic value and provisioning services of multi-species swards. Grassland Sci Eur 17:35–46

Jactel H, Branco M, Duncker P, Gardiner B, Grodzki W, Langstrom B, Moreira F, Netherer S, Nicoll B, Orazio C, Piou D, Schelhaas M, Tojic K (2012) A multicriteria risk analysis to evaluate impacts of forest management alternatives on forest health in Europe. Ecol Sociol 17:52. https://doi.org/10.5751/ES-04897-170452

Jentsch A, Kreyling J, Beierkuhnlein C (2007) A new generation of climate-change experiments: events, not trends. Front Ecol Environ 5:365–374

Käyhkö N, Skånes H (2006) Change trajectories and key biotopes–Assessing landscape dynamics and sustainability. Landsc Urban Plan 75:300–321. https://doi.org/10.1016/j.landurbplan.2005.02.011

Kedziora A (2010) Landscape management practices for maintanance and enhancement of ecosystem services in a countryside. Ecohydrol Hydrobiol 10:133–152

Lavorel S, Colloff MJ, Mcintyre S, Doherty MD, Murphy HT, Metcalfe DJ, Dunlop M, Williams MJ, Wise RM, Williams KJ (2015) Ecological mechanisms underpinning climate adaptation services. Global Change Biol 21:12–31

Lovett JC, Marchant R, Taplin J, Kuper W (2005) The oldest rainforests in Africa: stability or resilience for survival and diversity? In: Purvis A, Gittleman JL, Brooks T (eds) Phylogeny and conservation. Cambridge University Press, Cambridge, pp 198–230

Malavasi M, Carboni M, Cutini M, Carranza ML, Acosta ATR (2014) Landscape fragmentation, land-use legacy and propagule pressure promote plant invasion on coastal dunes: a patch-based approach. Landsc Ecol 29:1541–1550

Mouillot F, Rambal S, Joffre R (2002) Simulating climate change impacts on fire frequency and vegetation dynamics in a Mediterranean-type ecosystem. Global Change Biol 8:423–437

Nordén B, Dahlberg A, Brandrud TE, Fritz Ö, Ejrnaes R, Ovaskainen O (2014) Effects of ecological continuity on species richness and composition in forests and woodlands: a review. Ecoscience 21:34–45. https://doi.org/10.2980/21-1-3667

Oliver TH, Morecroft MD (2014) Interactions between climate change and land use change on biodiversity: attribution problems, risks, and opportunities. WIREs Clim Change 5:317–335. https://doi.org/10.1002/wcc.271

Pechanec V, Purkyt J, Benc A, Nwaogu C, Štěrbová L, Cudlín P (2018) Modelling of the carbon sequestration and its prediction under climate change. Ecol Inform 47:50–54

Pitkanen TP, Kumpulainen J, Lehtinen J, Sihvonen M, Kayhko N (2016) Landscape history improves detection of marginal habitats on semi-natural grasslands. Sci Total Environ 539:359–369

Plieninger T (2006) Habitat loss, fragmentation, and alteration–quantifying the impact of land-use changes on a Spanish Dehesa landscape by use of aerial photography and GIS. Landscape Ecol 21:91–105

Polce C, Kunin WE, Biesmeijer JC, Dauber J, Phillips OL (2011) Alien and native plants show contrasting responses to climate and land use in Europe. Global Ecol Biogeogr 20:367–379

Rehounkova K, Prach K (2007) Spontaneous vegetation succession in gravel-sand pits: a potential for restoration. Restor Ecol 16:305–312

Reusch TBH, Ehlers A, Hämmerli A, Worm B (2005) Ecosystem recovery after climatic extremes enhanced by genotypic diversity. Proc Natl Acad Sci 102:2826–2831

Sala OE, Chapin FS III, Armesto JJ, Berlow E, Bloomfield J, Dirzo R, Huber-Sanwald E, Huenneke LF, Jackson RB, Kinzig A, Leemans R, Lodge DM, Mooney HA, Oesterheld M, Poff NL, Sykes MT, Walker BH, Walker M, Wall DH (2000) Global biodiversity scenarios for the year 2100. Science 287:1770–1774

Scherber C (2015) Insect responses to interacting global change drivers in managed ecosystems. Curr Opin Insect Sci 11:56–62. https://doi.org/10.1016/j.cois.2015.10.002

Seják J, Cudlín P (2010) On measuring the natural and environmental resource value and damages. Stud Ecol 4:53–68

Seják J, Dejmal I, Petříček V, Cudlín P, Míchal I, Černý K, Kučera T, Vyskot I, Strejček J, Cudlínová E, Cabrnoch J, Šindlar M, Prokopová M, Kovář J, Kupka M, Sčasný M, Šafařík M, Roušarová Š, Stejskal V, Zapletal J (2003) Valuation and prizing of biotopes of the Czech Republic. Czech Ecological Institute, Czech, p 422. (in Czech)

Seják J, Cudlín P, Petříček V, Prokopová M, Cudlín O, Holcová D, Kaprová K, Melichar J, Škarková P, Žákovská K, Birklen P (2018) Methodology of habitat valuation. Nature Conservation Agency of the Czech Republic, Prague, p 232. (in Czech)

Simkin SM, Allen EB, Bowman WD, Clark CM, Belnap J, Brooks ML, Cade BS, Collins SL, Geiser LH, Gilliam FS, Jovan SE (2016) Conditional vulnerability of plant diversity to atmospheric nitrogen deposition across the United States. Proc Natl Acad Sci 113:4086–4091

Skaloš J, Novotný M, Woitsch J, Zacharová J, Svoboda M, Křováková K, Romportl D, Keken Z (2015) What are the transitions of woodlands at the landscape level? Change trajectories of forest, non-forest and reclamation woody vegetation elements in a mining landscape in Northwestern Czech Republic. Appl Geogr 58:206–216

Smiraglia D, Ceccarelli T, Bajocco S, Salvati L, Perini L (2016) Linking trajectories of land change, land degradation processes and ecosystem services. Environ Res 147:590–600

Swift TL, Hannon SJ (2010) Critical thresholds associated with habitat loss: a review of the concepts, evidence, and applications. Biol Rev 85:35–53

Thomas CD, Cameron A, Green RE, Bakkenes M, Beaumont LJ, Collingham YC, Erasmus BFN, de Siqueira NF, Grainger A, Hannah L, Hughes L, Huntley B, van Jaarsveld AS, Midgley GF, Miles L, Ortega-Huerta MA, Peterson AT, Philips OL, Williams SE (2004) Extinction risk from climate change. Nature 427:145–148

Thuiller W (2007) Climate change and the ecologist. Nature 448:550–552. https://doi.org/10.1038/448550a

Travis JMJ (2003) Climate change and habitat destruction: a deadly anthropogenic cocktail. Proc Royal Soc B-Biol Sci 270:467–473

Van Eetvelde V, Antrop M (2009) Indicators for assessing changing landscape character of cultural landscapes in Flanders (Belgium). Land Use Policy 26:901–910. https://doi.org/10.1016/j.landusepol.2008.11.001

Walker B, Salt D (2006) Resilience thinking: sustaining ecosystems and people in a changing world. Island Press, Washington, p 14

Wästfelt A, Arnbergb W (2013) Local spatial context measurements used to explore the relationship between land cover and land use functions. Int J Appl Earth Obs 23:234–244

Williams DW, Liebhold AM (1995) Forest defoliators and climatic change: Potential changes in spatial distribution of outbreaks of western spruce budworrn and gypsy moth. Environ Entomol 24:1–9

Wu J, Wu T (2013) Ecological resilience as a foundation for urban design and sustainability. In: Pickett S, Cadenasso M, McGrath B (eds) Resilience in ecology and urban design, future city, vol 3. Springer, Dordrecht

Young TP (2000) Restoration ecology and conservation biology. Biol Conserv 92:73–83n

Agriculture in the European Union: Seven More Years of Environmental Austerity?

Alicia A. Epstein

Abstract At a time when scientific consensus suggests that urgent and ambitious action is needed to address the ecological impacts of production agriculture, not least for realising international climate and environmental objectives, the political will to address these challenges has become increasingly stifled by several converging factors. In the European context, years of economic austerity has been followed by dramatic events including the decision of the United Kingdom to leave the European Union in 2016, as well as the endorsement of nationalist and populist agendas across a growing number of Member States. Together, these and other shifts in the geopolitical winds have already had direct implications for the way that environmental objectives are prioritised and funded under the European Union's Common Agricultural Policy. And, as the European Union seeks to conclude the next round of policy reform (covering the programming period 2021–2027), the legislative proposals currently being negotiated suggest that 'environmental austerity' could continue well into the future. Against this background, this chapter considers the extent to which the Common Agricultural Policy may be expected to deliver the outcomes necessary to meaningfully address ecological decline and climate change in the European Union during the next seven-year programming cycle.

1 Introduction

Following renewed scientific and societal focus in the late 2000s, the ecological impacts of the global food system have continuously demanded increased attention in international debates on food security, environmental protection and climate change, among others (Willett et al. 2019, Springmann 2018). With regards to land-use, in particular, key insights into the ecological impacts of production agriculture have been gained from the planetary boundaries framework, which provides a comprehensive account of the effects of human activity upon the Earth system and its

A. A. Epstein (✉)
Centre for Law and Administration, Brandenburg Technical University, Cottbus, Germany
e-mail: epstein@b-tu.de

L. Westra et al. (eds.), *Ecological Integrity in Science and Law*,
https://doi.org/10.1007/978-3-030-46259-8_17

sub-systems (Rockström et al. 2009). Not least, literature employing the planetary boundaries framework suggests that agriculture has been one of the main drivers of unsustainable resource use and has, consequently, contributed more heavily than most other sectors towards the transgression of planetary boundaries that define a safe operating space for humanity (Steffen et al. 2015). Thus, of the nine systems identified and analyzed by Rockström et al. (2009), they estimate that critical threshold boundaries have already been transgressed in at least three of these systems, all partly or mainly due to agricultural land-use and management practices.

However, while the effects of agricultural intensification have clearly had a global impact, the authors emphasize that much of the observed ecological deterioration has been caused by the current and historical actions of a few developed nations, with European Union (EU) agriculture certainly standing out as one of the main sources of this development (Stoate et al. 2009). Indeed, decades of continuous yield growth in Europe has been enabled by the employment of highly intensive management and production practices, including the expansion of monocultures and use of inputs such as pesticide and fertilizers, which have not only contributed to the disruption of the Earth's phosphorous and nitrogen cycles (Steffen et al. 2015), but have also had devastating impacts on terrestrial and marine biodiversity (Emmerson et al. 2016). As key sources of this decline, European and other economically advanced nations correspondingly bear a greater responsibility to reduce the impacts of their food sectors, than that of less developed states. Recent findings suggest that important international targets relating to e.g. emission reduction, the halting of biodiversity loss and sustainable water use cannot be realized without significantly reducing the global environmental footprint of agriculture (O'Neill et al. 2018). In the European context, it is therefore imperative that efforts to address ecological decline and climate change are accompanied by ambitious action in the field of agriculture.

Attaining the types of changes necessary to meaningfully reduce the environmental pressures stemming from EU agriculture will require major transformation of its Common Agricultural Policy (CAP). Yet, at a time when ambitious action is required to steer the CAP on a new course, the ability of the EU to take decisive steps towards this end has been stifled by a number of converging political developments in recent years. For instance, the decision by the UK to leave the EU following a 2016 referendum has since overshadowed most other political issues and resulted in a major financial shortfall that makes increased funding for environmental measures unlikely to be prioritized during the 2021–2027 budgetary period (Matthews 2016). Along with growing EU-skepticism and nationalism in other MSs, it would appear that the EU's ability to take bold and transformative action to address the ecological pressures stemming from agricultural, has consequently been limited by the risk of further alienating citizens and public opinion. Thus, as the EU prepares to reform the CAP for the next seven-yearly programming cycle (2021–2027), significant obstacles exist to meeting the level of ambition needed to address ecological and climate-related challenges.

Against this background, this chapter analyses the extent to which the CAP is prepared to deliver the outcomes needed to meaningfully address ecological decline and climate change in the EU. In particular, it questions the adequacy of current and proposed responses, given the scale and urgency of the challenges. The chapter proceeds in three parts, with the first part considering the role of the CAP in addressing ecological challenges in the EU. Secondly, the discussion will focus on the main elements and outcomes of the current legal framework before, thirdly, turning its attention to the reform proposals that are currently being negotiated.

2 Responding to Growing Ecological Challenges in the European Union: The Role of the Common Agricultural Policy?

With almost half (approximately 48%) of the territory under their management, farmers and other agricultural operators account for an extensive use of primary resources across the EU (IPCC 2019). The vast majority of these farmers receive subsidies, principally in the form of direct payments, which are financed by the EU and subject to the rules and conditions of the CAP. Given the share coverage of agricultural land, the CAP may therefore be distinguished as one of the EU's most environmentally impactful policies, having shaped the rural ecology of some Member States (MS) for over half a century (Jack 2016). In addition, the CAP has remained the single largest spending item in the EU budget since it was introduced in 1962. During much of this time, the financial strength of the CAP has been wielded to incentivise land use and production practices—through subsidies and other economic incentives—that have favoured intensive forms of production, with limited consideration for their long-term environmental implications (Emmerson et al. 2016).

As already emphasised in the global context, the unmatched increases in productivity that have been realized in recent decades, have likewise come at great environmental cost in the EU, where conventional agriculture has caused major disturbances to the ecological systems that enable food production (Springmann 2018). Consequently, the need to reduce the impacts of agriculture is widely recognized by the EU and has, since the early 1990s, mainly been pursued through the integration of environmental policy objectives under the CAP. However, despite decades of such policy integration and the development of an extensive agri-envrionmental framework, various indicators suggest that EU efforts have largely been unsuccessful in delivering the outcomes necessary to change the current trajectory and ensure a transition towards sustainable agriculture (Pe'er et al. 2014).

For instance, notwithstanding incremental improvements for the conservation status of a number of species in the last decade, the EU is not expected to hit its current headline target to halt biodiversity loss by 2020 (IPBES 2018). This would, thus, be the second time the EU has fallen short of reaching its own biodiversity goals, having also failed to meet the previous targets set under its 2010 biodiversity strategy (European Commission 2006). The sustained decline is particularly alarming given that biodiversity supports core services and functions, upon which the stability and integrity of ecological systems depends (Lohbeck et al. 2016). Moreover, persistent nitrate pollution of European waterways continues to pose major risks to the environment, as well as human and animal health in MSs such as the Netherlands, Denmark, Spain and Germany where low water quality has consistently been reported as a direct result of agricultural management and production practices in the livestock sectors (EU Commission 2018). Likewise, declining soil quality is a serious challenge across the EU, with 45% of soils exhibiting low or very low carbon content (Pe'er et al. 2014).

That said, some of the greatest challenges to future productivity in the EU are expected to result from climate change and increasingly disruptive weather patterns (IPCC 2019). Indeed, higher temperatures, irregular precipitation, rising sea levels and increased frequencies of drought have already been experienced across the EU, with varying regional implications for agriculture (Kovats et al. 2014). A particularly severe and recent example may be made with reference to the extensive drought experienced in the spring and summer of 2018 in Northern Europe, which resulted in reductions in main crop yields by up to 50%, compared to the previous year, across the affected regions (Reinermann et al. 2019). At the same time, it should be added that this drought coincided with especially wet conditions in parts of Southern Europe, including the Iberian Peninsula, where crop yields improved by up to 35% (Toreti 2019). Notwithstanding instances of increased yields, the EU's Joint Research Council expects that hotter and more extreme weather conditions will pose significant risks to food security in the long-term (Toreti 2019).

In addition to being one of the sectors most affected by climate change, agriculture is also a key driver of this process and is responsible for a significant share of EU greenhouse gas (GHG) emissions when spillover effects for other sectors are considered (Fellmann et al. 2018). On the contrary, as the main source of land-use, agriculture is in a unique position to contribute to climate change mitigation and adaptation strategies, for instance, by protecting soil carbon deposits and improving the capacity of soils to sequester and absorb atmospheric carbon (Fernández-Getino et al. 2017). Unlocking this potential is therefore seen as a key priority for addressing climate change and realizing emission reduction targets under the CAP (European Council 2014). In order to explore the extent to which such outcomes may be possible, it is necessary to consider key elements of the current CAP framework, as well as the EU Commissions pending proposals for future reform.

3 The 2014–2020 CAP Framework: Expected Outcomes

The CAP's agri-environmental framework was last updated in 2013, when the Council and European Parliament (EP) agreed on a reform package for the 2014–2020 period. As in the case of each major overhaul of the CAP, the 2013 reform ushered in multiple changes to the existing framework and of key concern, for present purposes, is the introduction of three so called 'greening' measures aimed at delivering benefits for biodiversity, as well as climate change mitigation under EU Regulation 1307/2013 (Direct Payment Regulation). The greening measures require farmers to observe EU-wide rules on crop diversification, the protection of permanent grasslands, as well as the establishment of Ecological Focus Areas, as a condition for receiving direct payments (income subsidies) during the 2014–2020 programming period (EU Regulation 1307/2013, Articles 43-47). In particular, the European Commission envisioned that they would 'ensure that all farms deliver environmental and climate benefits through the retention of soil carbon and grassland habitats associated with permanent pasture, the delivery of water and habitat protection…and the improvement of the resilience of soil and ecosystems through crop diversification' (2011a). The significance of the greening measures is further underscored by the decision to make 30% of direct payments (income subsidies) subject to their compliance.

Despite initial claims by the EU Commission regarding the potential for these measures to improve the environmental dividends of the CAP, implementation appears to have fallen short in a number of ways (Matthews 2013). For instance, implementation of the greening measures has been significantly restricted by exceptions to the general rules and minimum standards that were originally intended to have the broadest possible application (Hart and Menadue 2013). Thus, in 2018, 3 years after the greening measures entered into force, the JRC estimated that 71% of EU farmers remained unaffected by their introduction and incurred no compliance costs to this effect (Louhichi et al. 2018). For the 29% of farmers who *were* affected, the vast majority were only marginally so, with the Court of Auditors finding that, on average, the greening subsidies significantly exceed the cost to farmers of meeting the greening requirements (2017).

The high rate of farmers that are wholly or partially unaffected by the greening measures raises serious doubts about the level of ambition reflected by the current agri-environmental framework (Hauck et al. 2014). Indeed, environmental action groups have continuously warned that low thresholds for compliance, combined with broad exemptions from the general rules were destined to render the greening measures largely ineffective (Hart and Menadue 2013). They also levelled wide-scale criticism against the Commission for its design of the measures. In particular, the Institute for European Environmental Policy has questioned the choice by the Commission to introduce crop diversification, which requires farmers to grow a minimum number of crops relevant to the size of the holding, rather than crop *rotation*, which requires farmers to change the crops grown on each land parcel from one year to another, with evidence that the latter is likely to produce greater environmental benefits over time (Hart 2015). This was also recognised by the Commission in a 2011 Impact Assessment, which found that the introduction of crop diversification 'may not bring the full environmental benefits of crop rotation' (2011b). Despite this conclusion, the Commission nonetheless chose to support the adoption of the crop diversification measure on the basis that it would better fit the annual nature of CAP payments.

Together with other notable shortcomings, which are beyond the scope of the current discussion, the implementation of the greening measures has certainly left much to be desired in terms of their environmental impact. With regards to biodiversity, Pe'er et al., expect that the design and implementation of the current measures are 'unlikely to contribute to improving the status of farmland biodiversity given that the majority of farmers would not be required to perform any changes of current farming practices' in order to comply with them (2014). This conclusion is supported by Dicks *et al* who suggest that there is little empirical or experimental evidence to support the claim that wildlife may benefit from the greening measures (2014). Likewise, the ability of the greening measures to deliver benefits for the climate would appear to be comparatively limited under the current framework. Thus, in its 2016 Implementation Report the Commission estimated their contribution to be merely 'moderately relevant to address the priorities identified for GHG emissions, maintaining carbon stocks and/or increasing carbon sequestration' (European Commission 2016).

These outcomes have prompted the European Court of Auditors to conclude that the greening measures, as currently implemented, are unlikely to significantly enhance the CAP's environmental and climate performance (2017). On the other hand, it should be noted that the introduction of the greening measures has added significant complexity to the CAP in terms of both their administration and implementation (European Court of Auditors 2017). This burden is primarily borne by the MSs, where varying levels of technical expertise are expected to have further implications for the quality of implementation and enforcement (European Court of Auditors 2017).

4 Looking to the Future CAP: Seven More Years of Environmental Austerity?

As the EU prepares to reform the CAP for the 2021–2027 programming period, there is undoubtedly a growing need to improve the environmental dividends delivered by the CAP. And given the significant challenges outlined above, it is also clear that the ambition of future action will be required to far exceed that of current measures in order to produce meaningful outcomes. Accomplishing this high task will also require a level of commitment and determination, on the part of EU institutions and MSs, that goes well beyond efforts to date. Moreover, ambitious action to protect the environment will require appropriate funding allocations to this end. However, several current geopolitical issues threaten to undermine the EU's ability to secure considerable improvements during the 2021–2027 period. For the purposes of the current discussion two of these deserve brief mention.

First, it should be noted that all aspects of EU decision-making and governance have been affected by the decision of the UK to leave the EU, following a 2016 referendum. The CAP is no exception and stands to be impacted more than other policies as it loses funding from the UK, which has long been a net-contributor to the CAP budget (Matthews 2016). Consequently, the economic uncertainty brought about by the so called 'Brexit' is not likely to stimulate the EU's willingness to commit to major spending obligations in the near term.

Secondly, in addition to Brexit, the EU finds itself tested by the rise of nationalism and EU-skeptical politics in a growing number of MSs. Although this movement is not homogenous, it is defined by a readiness to challenge the EU on issues where over-reaching is perceived, as well as having a limited appetite to prioritize or support the enhancement of environmental protection (Matthews 2016). Moreover, following elections in May of 2019, newly elected Euro-skeptical MEPs have managed to form the third largest political group of the EP, thereby enhancing their power to collectively challenge EU policies at the institutional level (Ripoll 2019). These outcomes have also served to increase fragmentation within the EP, at the expense of groupings positioned along the traditional centre-right and centre-left divide, which have previously tended to be the dominant political forces. Given the legislative role of the EP, the composition and power of these groupings are expected to have a decisive impact on the type of reform that is agreed for the 2021–2027 programming period.

Against this background, it is suggested that the EU's ability to pursue ambitious environmental action in the field of agriculture may be significantly restrained by a desire not to alienate citizens or create further political divisions. Not least, the level of funding and ambition needed to transform agriculture is likely to be resisted by net-contributing MSs, while others could oppose such action on the basis that the EU is *imposing* its own will in the field of environmental protection against that of the MSs (Matthews 2016). In other words, if the EU is perceived to be 'running red lights' with its environmental agenda, further push-back is to be expected from anti-EU parties and politicians.

4.1 Key Elements of the Legislative Proposals

In June of 2018 the Commission adopted a package of legislative proposals to reform the CAP for the 2021–2027 programming period (European Commission 2018). These are currently subject to trilogue discussions between the Commission, the Council and the European Parliament (EP), which are likely to result in various changes being negotiated before the final measures are adopted under the ordinary legislative procedure by the EP and the Council. However, while the details of future instruments are difficult to gauge at present, there are several overriding themes that have defined the proposals and are expected to provide the basic framework of the reformed CAP envisioned by the Commission. Moreover, the legal texts have been accompanied by the publication of guiding policy documents and impact assessments that further set out the Commission's priorities and objectives for the CAP during the next programming period (European Commission 2017). For present purposes, two of the main themes of the proposals are likely to have implications for environmental protection and deserve further consideration.

First, the Commission has proposed the introduction of a new delivery model which would shift CAP governance from a compliance-based system to a performance—or outcome—based system. In doing so, the EU will be tasked with setting the basic policy parameters, objectives and requirements of the CAP, 'while Member States should bear greater responsibility as to how they meet the objectives and achieve targets', according to the recital to the proposed legislation (European Commission 2018). Each MS will be required to introduce a strategic plan for its entire territory (with possible regional implementation), setting out a comprehensive approach for implementing the new CAP framework. These must include a number of elements detailed in Articles 96-103 of the proposed legal texts and will, according to the Commission, ensure 'greater overall contribution to the achievement of the specific environmental-and climate-related objectives...in comparison to the overall contribution made to the achievement of the objective...in the period 2014 to 2020' (European Commission 2018).

Secondly, the new delivery model will be based on increased flexibility on the part of MSs, who will be able to choose from a range of options when implementing their strategic plans. Undoubtedly, this would mark a significant shift in the governance structure of the CAP, which has been defined by its highly centralised approach for decades (Henke et al. 2017). On the one hand, increased flexibility may enhance the ability of MSs to implement targeted measures that are more appropriate for local and regional conditions than centrally determined 'one-size fits all' measures. However, recent experiences relating to the implementation of the greening measures suggest that MSs generally 'do not use this flexibility to maximize the policy's environmental and climate benefits' (European Court of Auditors 2017). Thus, while increased flexibility is expected to unlock certain potentials, it is impossible to determine the extent to which MSs are likely to use their powers to simply choose those options that are most easily satisfied by their own farmers (Henke et al. 2017).

The Commission has tried to address these and other concerns by stating that national plans will have to reflect 'strong sustainability principles embedded in the CAP objectives' (European Commission 2019), but important questions undoubtedly remain as to the extent to which environmental improvements can actually be achieved under the future framework. Not least, the extension of increased powers and flexibility for MSs comes at a challenging time for the EU. Handing over more power to MSs to set their own objectives therefore also comes with considerable risks and few guarantees that the new framework will serve to enhance the environmental dividends of the CAP. Thus, until further details of the reforms are announced, it would appear that the current proposals will be unable to transform EU agriculture and deliver the outcomes necessary to reverse unsustainable resource use.

5 Conclusions

The monumental scale of the ecological and climate-related challenges posed to future productivity and agricultural systems, makes it imperative upon the EU to take the action necessary to provide meaningful and viable solutions. Indeed, given the considerable impact of EU agriculture, failure to adequately address these and other issues could contribute to further biophysical and biochemical tipping points being reached, with great risk to future food security. Consequently, it is difficult to overstate the need to create mechanisms that reflect the urgency of these challenges and produce outcomes capable of reversing the current trajectory. The need to act has perhaps never been greater.

As a flagship policy, it is inconceivable for the EU to reach its climate and environmental targets without also transforming the CAP. Yet, despite numerous reforms and the development of an extensive agri-environmental framework, the pace of change has been remarkably slow. The insufficiency of EU action was most recently demonstrated by the limited environmental ambition that defined the 2013 CAP reform, which delivered the policy's legal framework for the 2014–2020 budgetary cycle. Indeed, there would appear to be general consensus that they have failed to deliver the environmental benefits needed to meaningfully address the rate of ecological decline. And, as EU institutions engage in the final negotiations to reform the CAP for the next programming period, the question is whether the outlook is better for the next 7 years or if more of the same may be expected?

References

Dicks LV et al (2014) A transparent process for evidence-informed policy making. Conserv Lett 7(2):119–125

Emmerson M et al (2016) Chapter two – how agricultural intensification affects biodiversity and ecosystem services. Adv Ecol Res 55:43–97

EU Commission (2018) Report from the Commission to the Council and the European Parliament on the implementation of Council Directive 91/676/EEC concerning the protection of waters against pollution caused by nitrates from agricultural sources based on Member State reports for the period 2012–2015. COM(2018) 257

European Commission (2006) Halting the loss of biodiversity by 2010 – and beyond – sustaining ecosystem services for human well-being. COM(2006) 216

European Commission (2011a) Proposal for a regulation of the European Parliament and of the Council establishing rules for direct payments to farmers under support schemes within the framework of the common agricultural policy. COM(2011) 625

European Commission (2011b) Impact assessment

European Commission (2016) Mapping and analysis of the implementation of the CAP: 2016 implementation report. European Commission, Brussels

European Commission (2017) The future of food and farming. COM(2017) 650

European Commission (2018) Proposal for a regulation of the European parliament and of the council establishing rules on support for strategic plans to be drawn up by Member States under the Common agricultural policy (CAP Strategic Plans) and financed by the European Agricultural Guarantee Fund (EAGF) and by the European Agricultural Fund for Rural Development (EAFRD) and repealing Regulation (EU) No 1305/2013 of the European Parliament and of the Council and Regulation (EU) No 1307/2013 of the European Parliament and of the Council. COM/2018/392

European Commission (2019) Reflection paper: towards a sustainable Europe by 2030. COM(2019) 22

European Council (2014) Conclusions on 2030 climate and energy policy framework. European Council

European Court of Auditors (2017) Greening: a more complex income support scheme, not yet environmentally effective. European Court of Auditors Special Report: 21
Fellmann T et al (2018) Major challenges of integrating agriculture into climate change mitigation policy frameworks. Mitig Adapt Strat Glob Chang 23(3):451–468
Fernández-Getino AP et al (2017) Challenges and prospects in connectivity analysis in agricultural systems: actions to implement policies on land management and carbon storage at EU level. Land Use Policy 71:146–159
Hart K (2015) Green direct payments: implementation choices of nine member states and their environmental implications. IEEP, London
Hart K, Menadue H (2013) Equivalence mechanisms used for complying with greening requirements under the new common agricultural policy. IEEP, London
Hauck J et al (2014) Shades of greening: reviewing the impact of the new EU agricultural policy on ecosystem services. Change Adapt Socioecol Syst 1:51–62
Henke R et al (2017) The new common agricultural policy: how do member states respond to flexibility? J Common Market Stud 56(2):403–419
Intergovernmental Science-Policy Platform on Biodiversity and Ecosystem Services (2018) The regional assessment report on biodiversity and ecosystem services for Europe and Central Asia
IPCC (2019) Climate change and land: an IPCC special report on climate change, desertification, land degradation, sustainable land management, food security, and greenhouse gas fluxes in terrestrial ecosystems. UNEP, Nairobi
Jack B (2016) Agriculture and EU law. Routledge, New York
Kovats RS et al (2014) Europe. In: Climate change 2014: impacts, adaptation, and vulnerability. Part B: regional aspects. Contribution of Working Group II to the fifth assessment report of the intergovernmental panel on climate change. Cambridge University Press, Cambridge
Lohbeck M et al (2016) The importance of biodiversity and dominance for multiple ecosystem functions in a human-modified tropical landscape. Ecology 97(10):2772–2779
Louhichi K et al (2018) Economic impacts of the CAP greening: an application of an EU-wide farm model for CAP analysis. Eur Rev Agric Econ 45(2):205–238
Matthews A (2013) Greening CAP payments: a missed opportunity? IIEA, Dublin
Matthews A (2016) The potential implications of a Brexit for future EU agri-food policies. EuroChoices 15(2):17–23
O'Neill DW et al (2018) A good life for all within planetary boundaries. Nat Sustain 1:88–95
Pe'er G et al (2014) EU agricultural reform fails on biodiversity. Science 344(6188):1090–1092
Reinermann S et al (2019) The effect of droughts on vegetation condition in Germany: an analysis based on two decades of satellite earth observation time series and crop yield statistics. Remote Sens 11(15):1783
Ripoll A (2019) The new European Parliament: more eurosceptic? https://ukandeu.ac.uk/the-new-european-parliament-more-eurosceptic/. Accessed 15 Oct 2019
Rockström J et al (2009) A safe operating space for humanity. Nature 461:472–475
Springmann M (2018) Options for keeping the global food system within environmental limits. Nature 562:519–525
Steffen W et al (2015) Planetary boundaries: guiding human development on a changing planet. Science 347:1259855
Stoate C et al (2009) Ecological impacts of early 21st century agricultural change in Europe: a review. J Environ Manag 91(1):22–46
Toreti A (2019) The exceptional 2018 European water seesaw calls for action on adaptation. Earth's Future 7:652–663
Willett W et al (2019) Food in the anthropocene: the EAT–lancet commission on healthy diets from sustainable food systems. Lancet 393(10170):447–492. https://doi.org/10.1016/S0140-6736(18)31788-4

Part V
Ecological Integrity and the Geopolitical Situation

The Political Economy of Managing Without Growth

Peter Venton

Abstract Political economy includes the principles of the social sciences of economics, political science and sociology that apply to economic systems in political regimes. This chapter describes the political regimes of democracy and oligarchy in terms of their principles, institutions and purposes. Political economy is defined and analyzed to show that the implicit purpose of oligarchies requires a much higher rate of economic growth than democracies regire. Since World War II, Canadian governments established economic growth as a paramount policy objective. It is argued that this objective is central to a political regime of oligarchies and that Canadian governance has de facto been oligarchic. The results have been an excessive rate of economic growth, a substantial increase in economic inequality and degradation of the nation's environment all of which are inconsistent with the common good of democracy. Democratic outcomes can be achieved with a much lower rate of economic growth than has been the case in Canada especially since 1979. An example of policies of political economy for achieving democratic outcomes in Canada is taken from environmental economist Peter Victor's book *Managing Without Growth*. Victor shows how the rate of Canada's economic growth over a 30-year period could be reduced by 50% and at the same time, the incidence of poverty would be reduced by 50% while greenhouse gas emissions would be reduced by 31%.

1 Introduction

Political economy includes the principles of the social sciences of economics, political science and sociology that apply to economic systems in political regimes. This chapter describes the political regimes of democracy and oligarchy in terms of their principles, institutions and purposes. Political economy is defined and analyzed to show that the implicit purpose of oligarchy requires a much higher rate of economic growth than democracies. Since World War II, Canadian governments established

P. Venton (✉)
Canadian Peace Research Association, London, ON, Canada

L. Westra et al. (eds.), *Ecological Integrity in Science and Law*,
https://doi.org/10.1007/978-3-030-46259-8_18

economic growth as a paramount policy objective. It is argued that this objective is central to a political regime of oligarchy and that Canadian governance has de facto been oligarchic. The results have been an excessive rate of economic growth, a substantial increase in economic inequality and degradation of the nation's environment all of which are inconsistent with the common good of democracy. Democratic outcomes can be achieved with a much lower rate of economic growth than has been the case in Canada especially since 1979. An example of policies of political economy for achieving democratic outcomes in Canada is taken from environmental economist Peter Victor's book *Managing Without Growth*. Specifically, Victor shows how the rate of Canada's economic growth over a 30-year period could be reduced by 50% and at the same time, the incidence of poverty would be reduced by 50% while greenhouse gas emissions would be reduced by 31%. Victor shows that greenhouse gas emissions would be 60% lower than they would under a "business as usual" model projection of higher growth and other greenhouse gas policies that actually occurred in Canada over the period 1979–2006.

2 What Is Political Economy?

Political economy is an interdisciplinary study that draws upon the disciplines of economics, political science and sociology to analyze how economic systems are managed in political regimes of states to serve the interests of groups of their citizens (Wikipedia, Political Economy). These disciplines are social sciences in the sense that they employ the same scientific methods of figuring out the causality of empirical observation as are employed in the natural sciences. The difference is that they incorporate empirical social facts made by humans which are much different in kind from the facts of natural science. French philosopher Auguste Compte, who defined the term sociology in 1838, regarded sociology as the scientific understanding of social phenomena that could unify history, psychology and economics (Wikipedia, Sociology).

Economics is the study of the use of scarce resources to satisfy unlimited human wants (Lipsey et al. 1988, p. 5). It is about the instruments of efficient allocation without any specification of the value or purpose served by the satisfaction of those wants. Thus, like science, it is value free. Economic systems are about the production and distribution of goods and services among citizens and international trade and investment between states. Economics is central to analyzing economic systems but, while it is rational, it is narrow and limited because it excludes political and social considerations (Wikipedia, Political Economy).

Political science is about the theories, practices and facts of politics of governance that are about resource allocations through the tax expenditure system and government management of the economic system. Governance institutions include electoral systems, the legislature, executive and political parties. Much political science is derived from public choice and social choice theories. Public choice theory is about the use of economic tools to address traditional problems of political science that include the following (Wikipedia, Public Choice):

- voter participation or apathy and/or knowledge or ignorance about political and economic issues
- voter theories such as the median voter theory
- the benefit principles of linking taxes to public expenditures
- principal-agent problem of asymmetry of knowledge about political and economic issues between the public on the one hand and the government and the bureaucracy on the other
- political phenomena of economic rent seeking for special market privileges, special interest group lobbying, and institutional corruption
- influence of elections on policy choices about government deficits and public debt, and about economic policies related to central bank independence, monopolistic competition among firms in industries and industrial policy

Political science is also about foreign affairs and international governance over nation states and relations and resource allocations between nation states. Specific examples of institutions of governance are the United Nations Security Council, the World Health Organization, the World Bank, the World Trade Organization and the International Monetary Fund.

Social choice theory is about aggregating individual opinions, preferences and interests to reach collective decisions about what constitutes the social welfare of their society. Social choice blends elements of welfare economics and political science theories of voting. Using elements of formal logic for generality, analysis proceeds from a set of seemingly reasonable axioms of social choice to form a social welfare function or constitution (Wikipedia, Social Choice theory). In the case of democracy, the axioms that are voted upon represent the elements of the common good.

Sociology is about the statistical stratification of society into groups and the values, norms and cultures of those groups. (Wikipedia, Sociology). Values reflect the philosophy or the meaning of collective life in society which is related to the aggregations of the philosophies of individuals about the meaning of life in their society. As well politics is instrumental for making public choices that affect the elements of social welfare. In this regard, the identification of normative values in sociology are paramount over economics and politics. Thus, the values of sociology which are the outcome of the philosophies of life are paramount over economics and politics. Indeed Aristotle's view was that the primary purpose of Oikonomi (economics) was to give citizen householders the leisure for pursuing man's two chief ends, politics and philosophy (Newell 2011, p. 17). The statistics of sociology on values may serve as measures of the elements of social welfare for a society. They also analyze the role of culture, ethnicity and gender in explaining economic and political outcomes. Although they may be paramount over economics and politics, the reality appears to be that they are also shaped by economic systems and by politics. In this context philosophy is defined as the general principles or laws of a field of knowledge as in the philosophy of economics, politics and sociology and moral philosophy is imbedded in sociology.

What follows is a political economy analysis of the economic systems of capitalism and socialism in the political regimes of democracy and oligarchy. Political economy originated in moral philosophy in the eighteenth century and is attributed to British scholars Adam Smith, Thomas Malthus and David Ricardo. In the late nineteenth century the term economics gradually replaced political economy and thereby dropped the moral aspects in sociology and some of the public choice theory of political science (Wikipedia, Political Economy). Thus, I argue that political economy represents a more robust approach for analyzing political regimes.

3 Economic Systems

Two economic systems of capitalism and socialism are discussed in this chapter. In reality economic systems in advanced economies like the United States of America (USA) and Canada are a mix of capitalist and socialist systems. According to historian Niall Ferguson most young Americans do not know what socialism means (Ferguson 2019) and USA economist Joseph Stiglitz advises Democratic party politicians not to use the word socialist (Stiglitz 2019). The main generic differences between the economic systems among advanced economies in democratic regimes, like the USA and Canada, are the degree of the mix between the capitalist and socialist sectors and the difference between the ownership and regulation of business firms.

3.1 Capitalism

Capitalism is a system for organizing the economic affairs of society that is comprised of four interrelated institutions. Private capitalist firms engaged in the production and distribution of goods and services are the first of these. Markets for exchange are the second. These include markets for goods and services, markets for the exchange of labor services, markets for trading in financial instruments that include securities, bonds, mortgages, derivatives, and markets for the exchange of assets such as housing, buildings, equipment and consumer durables. The third institution is a monetary system based on bank credit, the purpose of which is to preserve the purchasing value of a currency in a nation state over time. The fourth institution is government coordination of the previous three institutions (Mann 2013, pp. 5,13).

There are many variations of capitalist economic systems depending on the extent and nature of government regulation of markets, regulation of private firms as well as taxation policies and expenditure programs and the ownership of the means of production such as public utilities for transportation and communications. The ultimate goal of capitalist firms is an ever-increasing accumulation of wealth by their owners and this translates into the goal of wealth of the nation state.

3.2 *Socialism*

According to the Oxford English Dictionary, socialism is a system of organization based on state or collective ownership and regulation of the means of production, distribution and exchange for the common benefit of all members of society (Ferguson 2019). It is similar to capitalism with the important difference that state ownership of firms is substituted for government regulation of firms. As well, the purpose of socialism is for the achievement of the common benefit of all members of society. This is the same purpose as that of democratic regimes.

3.3 *Mixed Capitalist/Socialism*

These systems are the same as capitalist economic systems with the added category of the broader public sector organizations for the production and distribution of goods and services outside of market systems. These include crown corporations, primary, secondary and postsecondary educational institutions, public transportation corporations, hospitals, federal, provincial and municipal governments and their agencies that provide a wide variety of services and products. As well, they include financial transfers for social services to individuals. In Canada, government sector spending accounts for almost 21% of GDP.

4 Democracy

The institutions and principles of representative democracy are described in Sect. 4.1 and the elements of the common good of democracy are outlined in Sect. 4.2. Section 4.3 identifies the main failure of representative democracy, as opposed to direct democracy, that has been evident for over 200 years.

4.1 *Institutions and Principles of Representative Democracy*

The following list is derived from (Venton 2016, pp. 192,193, 200–206) and (Venton 2017, pp. 248–250).

- A social contract balancing positive freedom of citizens to organize for the production and distribution of public services with negative freedom from responsibility for others through the tax expenditure system for providing public services
- A constitution and the rule of law about which all are informed that provides for free speech and the right of assembly and property rights
- Principles of equality, equal opportunity, and social justice

- Equal right to participate in the process of legislation
- Equal treatment of all groups
- Freedom to periodically elect political representatives who are expected to serve the interests of at least a majority of citizens
- A majoritarian electoral system
- Public education that informs citizens how to participate in deliberations about public issues
- Media institutions that keep the citizenry informed with news about the operation and policies of the legislature, the government and their political representatives

4.2 The Common Good of Democracy

The ultimate purpose of democracy in a nation state is the achievement of the "common good". The common good is about "social welfare" of the entire populace. It is a collection of positive and negative ends of society (Venton 2015, p. 208; Victor 2008, pp. 129–130).

Positive ends

- Living standards (measured by per capita GDP)
- Secure full employment
- Education levels
- Democratic engagement
- Peace and security
- Health
- Equality in the distribution of wealth and income
- Leisure time
- Quality of environmental resources

Negative ends

- Incidence of crime
- Incidence of family breakdown
- Pollution of air and water
- Inequality in the distribution of wealth and income
- Incidence of poverty

The optimal mix of the foregoing elements can be determined by sociological random surveys of public attitudes about the degree of importance that the "votes" of all who are surveyed attach to each element. The aggregation of these monetary values produces an arbitrary index number similar in kind to the statistical Consumer Price Index numbers that are used to measure the degree of inflation in prices of a representative basket of goods and services that are sold in the markets in the economy. Percentage changes in the resulting index numbers may be accepted as measures of progress for the entire society.

4.3 Failures of Representative Democracy

In the era of industrial capitalism, representative democracy has typically failed to represent the interests of the majority, The institution of representative democracy as it evolved in Europe and America in the nineteenth and twentieth century ruled in favor of the "sinister interests" of the holders of power at the expense of the interests of the whole community according to Jeremy Bentham (1815). John Stuart Mill (1859) also recognized this danger to democracy. The analysis of Bentham and Mill found echoes in the political philosophy and sociological analysis of Vilfredo Pareto (1916), Gaetano Mosca (1896), and the American C. S Mills (1956). Indeed C.S Mills identified a compact of economic, political and military elites who tended to think and behave in very similar ways that made a mockery of democracy (Bergandi 2017, pp. 181–184).

5 Oligarchy

5.1 Definition

Oligarchy is a political regime in which the wealthiest classes are disproportionately powerful and rule for the sake of their private interests (Neill 2011, p. 25). It is the opposite of democracy in the sense that it involves control by the few (rather than the many) and where the few are wealthy and the many are not wealthy (Simpson 2011, p. 74). In oligarchies, political participation and justice are unequal in contrast with equal justice and equal participation that define democracy.

Most oligarchies have constitutions that are governed by the rule of law. However, the administration of justice is unequal in oligarchic regimes. The preservation of unequal justice requires limitations on the freedom of speech where freedom of speech entails speech that is frank, open and without fear of recrimination and reveals and uncovers the truth as one sees it (Cooper 2011, pp. 200, 204).

The limitation on free speech requires a measure of deceit or sophistry on the part of oligarchic leaders. One of the elements of deceit is the use of ideology that is a cloak for deeper motivation for the excessive accumulation of wealth in the hands of a few or for the cause and unfolding of war against foreign states (Bagby 2011, pp. 122–123). Another example is found in George Orwell's 1984 speech using aphorisms like "war is peace", "freedom is slavery" and "ignorance is strength" which represent the power of holding two contradictory beliefs in one's mind simultaneously and accepting both of them (Bergandi 2017, pp. 179, 186). A recent example of duplicity is found in the term "sustainable development" that is sufficiently vague to allow multiple meanings which are in conflict. One meaning of the term implies protection of the environment for future generations while the term development allows for a continued adherence to economic growth that degrades the environment (Montini and Volpe 2017, p. 173).

Oligarchic leaders are not competent or trained to rule as they have a conception that blinds them to the fact that the realm of politics is different in kind from the realm of finance (Skultety 2011, p. 100). I have argued that effective leadership in democracies requires knowledge of all three of the elements of political economy, namely economics, political science and sociology and their underlying philosophies (Venton 2015, 2016, 2017). These are well beyond the paradigms of economics and finance.

The measure of success of the oligarchic nation is ever increasing national wealth. This success is achieved by maximizing Gross Domestic Product (GDP) each year, saving some it from consumption, and real productive assets and using the rest to increase financial wealth. Unlike democracy, there is no good purpose to this success because wealth leaves unanswered the purpose for which it is being amassed. The political objective of the oligarchic regime is good order which is defined by the temperament of the oligarchic leaders which tends to favor the accumulation of their private wealth which is included in the wealth of the nation.

In America today, oligarchy prevails in an informal way which differs from the ancient way of formal structure. The result is the same, namely that of a self-interested elite who are obstructing or operating contrary to the general welfare [i.e., the common good]. Broadly two approaches of the oligarchic elite are to use their money to influence elections and to shape the opinion of the public (Neill 2011, p. 34).

5.2 *Oligarchy and Neoliberal Capitalism*

In his book, *Can Democracy Survive Global Capitalism?*, Robert Kuttner traces the decline from a decent economy in America of the 1960s to its adoption of neoliberal capitalism that began in the 1980s under President Ronald Reagan's administration (Kuttner 2018). Canada was not far behind. The Canadian Federal government's massive austerity measures in the middle 1990s to balance the budget, reflected what political scientist Matt Fodor described as a solidification of the neoliberal counter revolution in Canada (Fodor 2013, pp. 109–112). However, the antecedents of this austerity began 25 years earlier around 1970. After 1970, Canadian federal and provincial governments reduced the very high tax rates that had been introduced to finance World War II but had been maintained for 25 years following the war. The reductions after 1970 included:

- A 59% reduction in federal corporate income tax rates from 37% in 1970 to 15% in 2012
- Repeal of the federal estate tax on inheritances in 1971
- An average 54% reduction in the 1969 highest marginal income tax rates for the highest 1% of income earners
- An average 32% reduction in the 1969 highest marginal income tax rates for middle income earners

Eventually these tax reductions took their toll in increased national debt and large annual deficits which precipitated major austerity initiatives in the mid-1990s by both the federal and provincial governments to balance budgets. Political leaders were informed by neoliberal leaders in America and Britain that tax increases to reduce deficits were not an option. Other neoliberal policies of the 1980s in America included increasing government subsidies to businesses, increasing demand for business products and services by means of privatizing government operations, reducing regulations on businesses and reducing regulations on financial institutions. These, together with tax reductions, represent what Thomas Piketty describes as the "forces of divergence" meaning the forces that concentrate wealth in the hands of a few citizens (Piketty 2014, pp. 21–22). They are the forces that are prominent in oligarchic regimes.

6 A Short History of Economic Growth

6.1 How Economic Growth Became a Paramount Policy Objective

Around the end of World War II, policies of full employment were adopted by several governments of the developed economies in the West. These included Britain's wartime government, the Australian Government, the Canadian Government and the government of the USA. Following John Maynard Keynes's analysis in the 1930s, Britain's Roy Harrod and E Domar of the USA developed the Harrod Domar model of economic conditions for generating sufficient economic growth to achieve full employment in 1946. Starting around 1950, economic growth was identified as a government policy. By the end of the 1950s economic growth had been thrust to the top as apparently an overriding objective of government policy. Economic growth was a major campaign issue in the contest between Kennedy and Nixon in 1960 after which faster economic growth became a central objective. By the end of the 1960s, the case for economic growth as an over arching policy objective in developed countries around the world had matured. Economic growth was apparently popular among political leaders as it came to be regarded as a remedy for all of the major current ailments of western economies: under employment, inflation, balance of payments difficulties, the dollar shortage and the Soviet challenge in the Cold War (Victor 2008, pp. 13–14).

At the same time as economic growth was reaching the pinnacle of policy objectives in the 1960s, several prominent economists began to question the validity of its measure, per capita GDP, as an indicator of social welfare or societal progress. These included the following and their authored works (Victor 2008, p, 15):

John Kenneth Galbraith, *The Affluent Society* (1958)
Kenneth Boulding, "The Economics of the Coming Spaceship Earth" (1966)
Ezra Mishan, *The Costs of Economic Growth* (1967)

E.F. Schumacher, *Small is Beautiful* (1973)
Herman Daly, *Steady State Economics* (1977)

Nevertheless, economic growth has remained a paramount objective to this day, despite its apparent inconsistency with the UN Sustainable Development Goals related to the 2030 Agenda for Sustainable Development. However, the inconsistency is only apparent because an explicit definition of the sustainable development concept in the UN Agenda is still lacking (Montini and Volpe 2017, p. 173). By default, the ambiguous definition of sustainable development in the 1987 Bruntland report has helped make it possible for governments, businesses and others to adopt the goal of sustainable development without compromising their adherence to economic growth (Victor 2008, p. 19).

6.2 Why Neoliberal Capitalism Needs Economic Growth

The reality is that the economic system of laissez fair capitalism favored by oligarchic regimes needs economic growth in order to provide full employment and hence give it political legitimacy for the majority of the populace. Profit rates in an unregulated capitalist economic system that is dominated by large corporations in monopolistic competition are too high. Consequently, only a fraction of profits are reinvested in the real productive capital in the economy. The rest are invested in financial assets in the financial sector. If all profits were reinvested, capitalists would have excess productive capacity that could not be sold and their profit rates would fall. In these circumstances, corporate capitalists need economic growth in demand for goods and services to utilize this excess productive capacity. That is one reason why capitalists are promoters of free trade policies that enable them to utilize some of their productive capacity to sell products abroad. It is also why they favour the liberalization of international investment that will enable them to invest some of their excess financial wealth abroad for more profits. Without external markets their options are limited to speculative investments in financial assets in domestic markets or low returns on safe investments.

In a growing economy, where the rate of return on capital exceeds the growth in national income, private wealth becomes more concentrated in the hands of a few (Piketty 2014). This phenomenon arises because some of the income from their wealth (i.e., return on their capital that includes interest on bonds, capital gains and dividends from securities, interest from mortgage repayments, and capital gains on their principal residences) is saved and reinvested in more financial assets that increase their financial wealth. The middle class, who make up the largest segment of voters, may not care as long as their incomes keep rising relative to their cost of living. It takes a considerable time of a few decades for the general public to realize that their employment incomes have barely kept pace with general inflation, or perhaps have declined in real terms, while the rich have gotten much richer. Indeed, "virtually all of the pre-income tax gains in Canada between 1982 and 2010 have gone to the top 10% of income earners (Banting and Myles 2016, p. 5)." Further

"The OECD estimates that the top 1% of Canada's income earners have captured 37% of total income growth over the past three decades (Banting and Myles 2016, p. 1). After almost 30 years, Canadians woke up to this kind of inequality following the Occupy Wall Street movement in 2011 that protested against the huge incomes of the top 1%.

6.3 Disappointments with Economic Growth

Since the 1980s, economic growth has been a disappointment. "Despite a 109.5% growth in real GDP, Canada's rate of unemployment barely fell from 7.5% in 1979 to 6.3% in 2006 (Victor 2008, p. 157).

Growth in real GDP did not reduce poverty significantly as measured by the Human Poverty Index (Victor 2008, p. 160). The income gap of an average Canadian family living below the Low Income Cut Off (LICO) poverty line was greater than it had been 25 years previously even though the real economy had grown by 100% (Victor 2008, p. 160).

Economic growth has been accompanied by a major increase in the inequality of income. "The share of total pre-tax income (excluding capital gains) received by the top one percent of income earners increased from about 7.5 per cent in the late 1970s to 13.5% in 2000 while the share of the rest of the 10 per cent hardly changed. Meanwhile, the share of total income going to families with middle incomes (defined as between 75 per cent and 150 percent of median after-tax income) fell from 52.1 percent in 1989 to 47.3 percent in 2004 (Victor 2008, p. 163)."

Last, but not least, economic growth has increased green house gas emissions thereby increasing the prospect of a "terror of an unliveable future" for our grandchildren (Klein 2014, p. 28). In addition, economic growth has been generating material and energy flows that are increasing beyond the capacity of the environment to accommodate them (Victor 2008, p. 191). More specifically, economic growth involves an increase in "throughput" which is the flow of natural resources from the environment through the economy and back to the environment as waste (Montini and Volpe 2017, p. 173). Therefore, slower economic growth is needed to reduce green house gas emissions and eliminate excessive throughput to preserve the environmental system for future generations.

7 Conclusions About Peter Victor's *Managing Without Growth*

In his 2008 book, *Managing Without Growth: Slower by Design, Not Disaster,* environmental studies professor Peter Victor presents a set of policies based on a sociological, political science and economics analysis (i.e., political economy) for reducing green house gas emissions through slower economic growth in Canada over a hypothetical projected 30-year period from 2005 to 2035.

7.1 Policies

- Reduce the work week by 15% for more leisure and less consumption.
- Ensure that increased labor productivity is taken up with leisure time rather than output and consumption.
- Eliminate poverty through tax changes to redistribute after-tax income.
- Increase the capital tax rate to favor investment in people over structures and machinery.
- Invest more in public infrastructure to reduce crowding.
- Invest less in the production of private status goods.
- Apply differential tax rates on status goods which otherwise spur more consumption.
- Invest more in wasteful deferred maintenance.
- Tax economic rent of big monopolistic businesses.
- Introduce a tax on carbon and regulate carbon emissions.
- Limit immigration of economic migrants to lower the rate of population growth.

7.2 Projected Results

Victor employs an econometric model to project what would happen as a result of his proposed policies. The results presented below are for the sixth of his six scenario projections over a 30 year projection period from 2005 to 2035 (Victor 2008, p. 177, Table 10.1). The first of his six projections is a "business as usual" scenario based on the assumption that the Canadian economy would perform, on average, much the same way as it did over the period from 1979 to 2006.

- The average annual increase in per capita GDP is reduced from 2.6% to 1.3%.
- Greenhouse gas emissions are reduced by 60% compared with the business as usual model and by 31% from the base year of 2005.
- The UN Human Poverty Index is reduced by 54% from a rate of 10.7% to a rate of 4.9%.
- The unemployment rate is reduced by 45% from a 2005 rate of 7.2% to a rate of 4.0%.
- Government sector expenditures are increased from 21.7% of GDP to 24.4% thereby reversing the decline since the 1990s.
- The ratio of government sector debt to annual GDP is reduced from 62% in the base year 2005 to 30% in 2035.

References

Bagby L (2011) Thucydides and the importance of ideology in conflict. In: Tabachnick D, Koivukoski T (eds) On oligarchy: ancient lessons for global politics. University of Toronto Press, Toronto, pp 110–139

Banting K, Myles J (2016) Framing the new inequality: the politics of income redistribution in Canada. In: Income inequality: the Canadian story. Institute for Research in Public Policy, Montreal. February 2016

Bergandi D (2017) The ecological catastrophe: the political-economic caste as the origin and cause of environmental destruction and the pre-announced democratic disaster. In: Westra L, Gray J, Gottwald F-J (eds) The role of integrity in the governance of the commons: governance, ecology, law, ethics. Springer Publishing AG, New York, pp 179–189

Cooper C (2011) Oligarchy and the rule of law. In: Tabachnick D, Koivukoski T (eds) On oligarchy: ancient lessons for global politics. University of Toronto Press, Toronto, pp 196–201

Ferguson N (2019) Young Americans need to learn what socialism actually means. Globe and Mail, May 18, 2019

Fodor M (2013) Taxation and neo-liberal counter-revolution: the Canadian case. In: Himelfarb A, Himelfarb J (eds) Tax is not a four letter word: a different take on taxes in Canada. Wilfrid Laurier University Press, Waterloo, pp 101–117

Klein N (2014) This changes everything: capitalism versus the climate. Alfred Knopf Canada, Toronto

Kuttner R (2018) Can democracy survive global capitalism? W.W. Norton, New York

Lipsey R, Purvis D, Steiner O (1988) Sixth Edition economics. Harper & Row, New York

Mann G (2013) Disassembly required: a field guide to actually existing capitalism. AK Press, Oakland

Montini M, Volpe F (2017) Sustainable development: renaissance or sunset Boulevard? In: Westra L, Gray J, Gottwald F-J (eds) The role of integrity in the governance of the commons: governance, ecology, law, ethics. Springer Publishing AG, New York

Neill J (2011) Aristotle and American oligarchy: a study in political influence. In: Tabachnick D, Koivukoski T (eds) On oligarchy: ancient lessons for global politics. University of Toronto Press, Toronto, pp 24–46

Newell J (2011) Oligarchy and *Oikonomia*: Aristotle's ambivalent assessment of private property. In: Tabachnick D, Koivukoski T (eds) On oligarchy: ancient lessons for global politics. University of Toronto Press, Toronto, pp 3–23

Piketty T (2014) Capital in the twenty first century. English Translation by Arthur Goldhammer. The Bellknap Press of Harvard University Press, Cambridge

Simpson P (2011) A corruption of oligarchs. In: Tabachnick D, Koivukoski T (eds) On oligarchy: ancient lessons for global politics. University of Toronto Press, Toronto, pp 70–89

Skultety S (2011) The threat of misguided elites: Aristotle and oligarchy. In: Tabachnick D, Koivukoski T (eds) On oligarchy: ancient lessons for global politics. University of Toronto Press, Toronto, pp 90–109

Stiglitz J (2019) An icon of the left tells democrats: don't go socialist. Foreign Policy.Com interview of Joseph Stiglitz May 20, 2019. http://foreignpolicy.com/2019/05/20/an-icon-of-the-left-tells-democrats-dont-go-socialist/. Accessed 15 June 2019

Venton P (2015) Radical changes in Canadian democracy: for ecology and the 'public good'. In: Westra L, Gray J, Karageorgou V (eds) Ecological systems integrity: governance, law and human rights. Routledge, New York, pp 201–219

Venton P (2016) Democratic capitalism for realizing the Earth Charter Vision. In: Westra L, Gray J, D'Aloia A (eds) The common good and ecological integrity: human rights and the support of life. Routledge, New York, pp 192–210

Venton P (2017) Pope Francis's ethics for democratic capitalism and the common good. In: Westra L, Gray J, Gottwald F-J (eds) The role of integrity in the governance of the commons: governance, ecology, law, ethics. Springer Publishing AG, New York, pp 237–253

Victor P (2008) Managing without growth: slower by design, not disaster. Edward Elgar Publishing, Northampton
Wikipedia, Political Economy. https://en.wikipedia.org/wiki/PoliticalEonomy. Accessed 12 June 2019
Wikipedia, Public Choice. https://en.wikipedia.org/wiki/Public_choice. Accessed 27 Apr 2014
Wikipedia, Social Choice. https://en.wikipedia.org/Social-choic-theory. Accessed 16 Oct 2019
Wikipedia, Sociology. https://en.wikipedia.org/wiki/Sociology. Accessed 11 Oct 2019

Opportunity Within Failure: Can the Global Pact for the Environment Learn from Responsibility to Protect?

Prue Taylor

Abstract The Global Pact for the Environment is a legal initiative that seeks to promote significant change to international environmental law and governance. Initial efforts to gain state acceptance through United Nations processes have faltered. This chapter considers what can be learnt from the emergence and acceptance of the doctrine of 'responsibility to protect', that could inform future efforts to pursue either the Global Pact or other more transformative legal initiatives. The main finding is that the history of 'responsibility to protect' has much offer civil society in the search for transformative approaches to state sovereignty in an age of collapsing ecological integrity.

1 Introduction

The Global Pact for the Environment (GPE) came to international prominence very quickly. In early 2017 the President of the French Constitutional Council invited a group of legal experts to create a draft document. A few months later, a working draft was publicly presented to the world, under the auspices of the French legal think tank, *Le Club des Juristes*. By May 2018 the United Nations General Assembly (UNGA) had passed a resolution establishing an ad hoc open-ended working group (OEWG) to begin consultations on the GPE and calling for a technical report to guide that work. Following completion of the technical report (known as the 'gap report'), the OEWG met three times in Nairobi (January, March and May 2019) to consider the report and recommend next steps to the UNGA (GPE.org). However, by the end of May 2019, GPE proponents declared the final recommendations a 'failure' and a 'serious set-back' for the GPE. Other rhetoric called the outcome

P. Taylor (✉)
School of Architecture and Planning, University of Auckland, Auckland, New Zealand
e-mail: Prue.taylor@auckland.ac.nz

L. Westra et al. (eds.), *Ecological Integrity in Science and Law*,
https://doi.org/10.1007/978-3-030-46259-8_19

'very disappointing', 'weak' and a 'clear retreat from the original proposal' (GPE Newsletter, 27 May 2019; ENB 2019c, pp. 3 and 11).

What are we to make of this failure? Specifically, what are we to make of such failures at a time of clear crisis for international environmental governance? Is it possible that this 'failure' may be helpful if we can make positive use of the opportunities created? This framing is not intended to disrespect the hard work of GPE proponents, OEWG participants and many others. Rather the intention is to utilise the GPE experience in a positive way. Much academic analysis to date has focused on renovating the existing draft GPE or on replacing it altogether with a document more fit for purpose (Biniaz 2019; Kotzé and French 2018; Kotzé and Kim 2019). The focus here is different. This chapter asks; what can we learn from the emergence and state acceptance of other transformative concepts of international law that might be useful for (a) a reformed GPE (or alternative document) or (b) other transformative legal initiatives emerging from civil society such as Earth Trusteeship (Bosselmann 2015), Ecological Law (Oslo 2016) and Earth System Law (Kotzé and Kim 2019; Fernández and Malwé 2019)?

More specifically, what can we learn from the emergence and acceptance of a radical legal norm which emerged some 15 years ago and transformed state sovereignty and international law on humanitarian intervention? This legal norm is known as the 'responsibility to protect' doctrine (R2P). This doctrine could be relevant to international environmental law and governance in a number of significant ways. For example, it may be extended to apply to grave ecological harm within state jurisdiction. This was part of R2P's original scope (Report 2001, art 4.20). Support for this approach may have strengthened in recent years due to enhanced legal recognition of environmental human rights (Axworthy and Rock 2009; Fishel 2018). Another approach could be to analyse and utilize R2P's original reconceptualization of 'sovereignty as responsibility' to reform permanent sovereignty over natural resources (Deng 2010). These approaches all merit further investigation but they are not the focus of this chapter which is primarily on the process by which R2P emerged and what lessons this might hold for the GPE or other civil society initiatives.

This chapter begins with an overview of the GPE; what it is, its ambition and brief consideration of why it failed. It then uses selected factors central to the success of R2P and considers the extent to which these are present or absent in the case of the GPE. This preliminary analysis indicates that more in depth research is merited. Much can be learnt from a process that was more bottom up, utilised very broad consultation processes to build support and legitimacy, embraced normative change and was explicitly embedded in both European and non-European legal systems. A further conclusion is that all future transformative legal initiatives must pay much more attention to the processes by which legal initiatives evolve and develop. Failure to do so puts the legitimacy of international law at risk.

2 Overview of the Global Pact for the Environment

The draft GPE is a short document comprising a preamble which ends with the phrase "[a]ffirming the need to adopt a common position and principles that will inspire and guide the efforts of all to protect and preserve the environment [the Parties have] agreed as follows:" (GPE 2017). The GPE then lists twenty-six principles of which twenty could be considered basic principles of international environmental law. At first glance it appears to be a brief and generalized synthesis of the 1992 Rio Declaration, 1982 World Charter for Nature, the Earth Charter and other international environmental legal instruments.

The ambition or objectives of the GPE can be described as both political/aspirational and as being of a more technical legal nature. As regards the first, Laurent Fabius referred to the GPE as a response to the 'post Paris paradox' i.e., the growing urgency to act on climate change in face of the very significant loss of state political will to act (Fabius 2017).[1] Fabius chaired initial GPE drafting sessions, was the President of Climate Change COP 21 (which delivered the lauded 2015 Paris Agreement) and is the current President of the French Constitutional Council. In 2017 he became a United Nations (UN) Environment Patron on Environmental Governance in recognition of his political expertise in achieving the Paris Agreement and his initiation and promotion of the GPE (Unenvironment.org). Laurent Fabius is clearly the legal and political figurehead of a French diplomatic and legal initiative intended to strengthen global support for universal action to protect the environment.[2]

From a more technical legal perspective, the GPE's objectives are threefold: (1) to be a global legally binding environmental law instrument, based on the consensus agreement of states; (2) to clarify and codify all major principles of international environmental law in one instrument; (3) and progressively develop the law by assisting the evolution of principles from soft to hard law and achieving global recognition of a human right to live in an ecologically sound environment. To this end, the GPE includes well recognised principles of law (precaution, prevention and sustainable development), together with some emergent principles (non-regression, role of non-state and subnational actors) (Kotzé and French 2018). Leaders of the GPE's drafting identify seven current deficiencies in the body of international environmental law which justify the need for the Pact and have shaped its content and character as an 'overarching statement of binding principles' (Aguila and Viñuales 2019a).

Why then did the Pact's main proponents declare that it failed and suffered a serious set-back, following the OEWG's final meeting in May 2019? To at least begin to unravel this we need to first define what failure means in this context. It is used here to mean the difference between the OEWG's final recommendations to the UNGA and what the proponents desired.

[1] This is my interpretation of Fabius' phrase "significant concerns about future decisions".

[2] The additional involvement of the French President and government is considered further below.

The most important OEWG recommendation was that the UNGA: "forwards these [general] recommendations to the UNEA for its consideration, and to prepare, at its fifth session in February 2021, a political declaration for a UN high-level meeting, … in the context of [the 50 year commemoration of the 1972 UN Conference on Human Environment to be held in 2022], with a view to strengthening the implementation of IEL, and international environmental governance …" (OEWG Report 2019, 55(b)). This was a clear setback from the original proposals both in terms of ambition and timing. Regarding ambition, the recommendation merely refers to a vague and simple '*political declaration*'. This is a clear retreat from the objective of a *legally binding international instrument* that enshrines general principles of environmental law. There is no reference to the GPE (or any instrument of its nature) in the final recommendation. Nor is there any reference to the initiation of intergovernmental negotiations, but merely a UN high-level meeting. Regarding timing, a political declaration might be expected in 2022. The proponents aimed for adoption of the GPE (or similar instrument) by 2020–2021. In sum, there was no mandate for GPE negotiations to begin or for any instrument that would "constitute a step further than the [soft law] Rio Declaration." (Aguila and Viñuales 2019a, p. 9).

From a longer-term perspective, resort to a weak political declaration by 2022 may represent more of a temporary (albeit serious) setback for GPE proponents. The consensus outcome has been interpreted by a OEWG co-chairs as at least providing a 'first step in a continuing process.'(ENB 2019c, p. 1). The UNGA President's engagement in the process was also interpreted as providing "hope that the doors will remain open to rescue the initial ambition of the pact" if articulated as a contribution to the 2030 Agenda for Sustainable Development (ENB 2019c, p. 11; Tigre 2019).

Why then did the OEWG deliver such a weak outcome? A full analysis of the reasons is beyond the purpose of this chapter. However, it will briefly canvas a few theories including; the need for a consensus outcome, the influence of the 'gap report', and the lack of innovation and normative ambition.

As noted above, the OEWG recommendations to the UNGA represented the best that a consensus process could deliver. The proponents stated: "[p]art of the explanation for … failure lies in the so-called "consensus" method, which allows a small minority of countries (in this case, especially the United States, Brazil and Russia) to block an initiative that is desired by a majority of States. It should be remembered that it was a majority of States that adopted the Resolution of the UN General Assembly on 10 May 2018 to initiate the Nairobi negotiations: 143 states in favour/5 states against." (GPE Newsletter 27 May 2019). This frustration was reinforced at a recent World Leaders Debate on the GPE. The UNGA President reflected on the serious limitations of current international law and governance and lead debate on "whether it is time to revise notions of consensus and unanimity in decision-making."(Tigre 2019, p. 2).

A few observations are warranted here. First, the need for the OEWG to reach a consensus outcome may well have been the legacy of a rushed process to gain the original UNGA resolution mandating the consultations (UNGA Res 72/277). This resolution was passed by recorded vote (rather than by adoption without vote) with

143 states in favour, 6 abstentions and 6 (not 5) against. While a majority of states were in favour, some notable states (USA and Russia) voted against. Furthermore, a close reading of states recorded concerns reveals early dissatisfaction with the processes adopted including a lack of consultation, concerns over sovereignty, inadequate consideration of views of states, text suggestions being ignored and the decision to begin a process toward the GPE being based on a vote, rather than consensus. Second, reports from the working group sessions, particularly the third session, reveal a broad array of concerns being expressed by a wide spectrum of states. This suggests that Russia, Brazil and the USA were not entirely to blame as blockers of the GPE initiative. Finally, negotiations were affected by changes in government. Allies at the outset were no longer in favour and rejected a potential new agreement (Tigre 2019, p. 2).

Another possible factor in the OEWG outcome was the influence of the UN Secretary General's "technical and evidence-based" report on "possible gaps in international environmental law and environment related instruments" which guided OEWG discussions (UNGA Res 72/277). This became known as the 'gap report' (Gap Report 2018). The need for this report may well have emerged from state's concerns regarding UNGA Res 72/277, that they were not being listened to and that weaknesses in international environmental law resided elsewhere, including lack of implementation and inequities in global governance (Kotzé and French 2018; French and Kotzé 2019). Regardless of motivation, a commentator close to the report opined that it had a critical role in building political support. States needed to be convinced that the GPE 'adds value' (Voigt 2019). More fundamentally, the gap report became a case of form following function. In this case, that function became framed in a narrow, technical business as usual way.[3] Consultation reports reveal that this framing created a significant distraction for states and others, with much discussion around different understandings of what a 'gap' is or is not (ENB 2019a, b). In short, substantive discussions that could have built stronger support for the GPE in principle, did not receive sufficient attention. A related consideration was the gap report's potential chilling effect on the overall ambition of consultations based on a "negative … assumption that what States do not want to hear they should not in any case receive."(French and Kotzé 2019, p. 31; Biniaz 2019).

Finally, a more substantive problem may well have been the Pact's lack of innovation and normative ambition, creating a sense that it was not at all 'fit for purpose'. In other words, it was an inadequate response for the global ecological crisis confronting life on Earth. In a very insightful and thorough analysis Kotzé and French interrogate the text from a variety of relevant perspectives including the emergence of ecological integrity as a fundamental *grundnorm*, eco-centric ethical paradigms, earth system science, and environmental/nature's rights and find it inadequate. To take but one example of particular importance to this book, while the text includes some references to 'ecological integrity' overall its use is weak, confusing

[3] Voigt states that the report attempted to avoid this narrow framing by adopting a wide interpretation of its mandate. Some states suggested that report recommendations went beyond the UNGA mandate (Voigt 2019, p. 17).

and inconsistent. Article 18 is illustrative "In order to conserve, protect and restore the integrity of the Earth's ecosystem and community of life," parties are merely required to "co-operate in good faith and in the spirit of global partnership for the implementation of the Pact."

In the opinion of Kotzé and French, "only if the Global Pact were to incorporate ambitious normative provisions to strengthen … global governance efforts that aim to halt the deterioration of Earth system integrity, as well as maintain and improve integrity, will it be able to offer a firm foundation …required to confront head on the deep socio-ecological crisis..." (French and Kotzé 2019, p. 811). As it fails to meet this standard, they consider it a *stillborn initiative*—one which does not provide for foundational change. Other academic commentary criticises the Pact's failure to utilise earth system science and (in particular) the planetary boundaries framework both of which are foundational to a very different understanding of the complex global crisis (Fernández and Malwé 2019; Kotzé and Kim 2019; CHH 2019). It is hard to determine what impact these omissions had on states during consultations. It was largely left to civil society to champion truly transformative initiatives at OWEG meetings (e.g.; CCH 2019; Earth Trusteeship 2019). However, the very absence of a relevant text gave states an easy pass reinforcing a view that no fundamental changes in state behaviour were required.

With this overview of the GPE in mind, this chapter now turns to consider whether an important opportunity has presented itself. The GPE was a fast moving civil society initiative developed outside formal UN processes. As we have seen, it suffered a serious set-back once it entered the UN system. This may or may not be a good thing, depending on one's view of its purpose and content. Nevertheless, if a reformed GPE or other transformative legal initiative arising from civil society is to avoid a similar fate, it is relevant to ask -what can we learn from legal precedents that were more successful? This question is posed assuming that the state-centred international legal system will remain unchanged, in the immediate future. One possible precedent is the responsibility to protect doctrine (R2P).

3 Responsibility to Protect Doctrine

R2P was chosen because it is considered to have radically transformed the pre-existing concept of state sovereignty, as applied to humanitarian intervention. In 2000, UNSG Kofi Annan posed this challenge: "if humanitarian intervention is indeed an *unacceptable assault on state sovereignty* how should we respond to a Rwanda, to a Srebrenica – to gross and systematic violations of human rights that offend every precept of our common humanity?" (Millennium Report 2000, p. 48). In the current context of massive ecological destruction within the jurisdiction of states (forest fires in Brazil, Bolivia and Indonesia being some of the most recent examples), we can restate the same fundamental challenge as follows: 'if a legal obligation of *responsibility for the Earth's ecological integrity is an unacceptable assault on state sovereignty*, how should we respond to the gross and systemic destruction of Earth's ecological integrity that violates every precept of our common humanity?'

What is the essence of R2P and how did it respond to the problem of state sovereignty? Traditionally, humanitarian intervention was treated as conflicting with the *sovereign right* of non-interference with the internal affairs of states. R2P moved beyond this *sovereign rights* paradigm and reconceptualised state sovereignty (in the context of humanitarian matters) as the *legal responsibility* of states to protect people within their territory from certain harms. Initial work on the scope of R2P included natural and environmental catastrophes causing loss of life as a trigger for R2P (ICISS 2001), however states subsequently limited its scope to genocide, war crimes, crimes against humanity and ethnic cleansing. This is considered the first pillar of R2P. The second pillar extends to the international community which has the responsibility to assist and encourage states to meet their responsibility to protect. If states cannot or will not protect people within their territory, the third pillar comes into play; an international process can legally mandate other states to exercise collective responsibility including intervening to protect people within another states territory (Thakur 2019 and Global Centre for R2P).

It must be acknowledged that many significant and important criticisms exist concerning key aspects of R2P, including its lack of proper implementation in recent humanitarian crises (e.g.; Libya in 2011 and Syria). However, the focus for the purposes of this chapter is on the fundamental normative change achieved by R2P and the international process by which this came about.

R2P has been described as representing a 'fundamental conceptual' or 'tectonic shift' in state sovereignty (Brunnée and Toope 2008, p. 128; Slaughter 2005, p. 627). Referring to a 2004 High Level panel report Slaughter comments: "In two scant paragraphs the panel redefines sovereignty itself for all UN member states. These paragraphs accomplish a tectonic shift, reinterpreting the very act of signing the [UN] Charter in ways that will create a new legal and diplomatic discourse about member states' obligations to their own people and to one another." (Slaughter 2005, p. 627). Gareth Evans, a co-author of the panel report highlights several key elements of this reframing and [notes]: (1) the language 'responsibility to protect' rather than 'right to intervene' required opponents to rethink their arguments to engage with the issue; (2) it increased the number of actors potentially responsible and redistributed responsibility. If a state could not protect people within its jurisdiction, the international community had a responsibility to assist. If this was not adequate then responsibility 'trips upwards' to the international community; and (3) prevention is a key element with the international community sharing responsibility to prevent humanitarian crises (Evans 2015).

In short, R2P institutes a critical normative shift which requires states to behave differently and it creates a system of international accountability to ensure that they do so (Thakur 2019; Deng 2010). Further, some consider "no idea has moved faster in the international normative arena" than R2P (Weiss and Kuele 2011, p. 1). Critically, it did not emerge via the sort of consensus treaty process followed by the GPE but was adopted through a very different route. What can we learn from this? The next section outlines selected factors considered central to the success of R2P and considers the extent to which these are present or absent in the instance of the GPE. This is a preliminary analysis intended to determine whether more in depth research is merited.

4 A Preliminary Comparison

4.1 A Crisis vs What Crisis?

The timing of R2P was critical and it was clearly intended to respond to international humanitarian crises. The horrors of Rwanda and Srebrenica, together with the muddled and dangerously self-serving intervention of the United Kingdom in Kosovo, resulting in NATO air strikes (Evans 2015) were shaking the international community. During this period the understanding of collective and national security was moving towards human security and recognition that states themselves could represent the largest threat to individuals within states (Thakur 2011).

In contrast, the GPE is strangely silent on the urgency, scale and underlying causes of the massive complex global ecological crisis confronting us. The Preamble references growing environmental threats, the urgency of action on climate change and unprecedented biodiversity loss. However, the articles mostly lapse into a shallow anthropocentric framing of 'environmental harm', with the environment being an ambient consideration rather than foundational for all life. The GPE does not convey an intention to urgently address the existential risk for life nor does it speak to the Greta Thunbergs of this world or the peoples, nations and ecosystems already suffering. This had an impact on OEWG consultations with an NGO representative and Guyana noting the lack of ambition, urgency and credibility. As a result, the Pact would make little contribution to the current trajectory of ecological degradation. The US, Saudi Arabia and others opposed wording 'swift action is needed' to address environmental challenges (ENB 2019c). This may change in the future evidenced by a key actor's comment that we need law responsive to the anger of today's youth (Tigre 2019).

4.2 Insiders vs Outsiders

The pathway towards R2P was primarily initiated by UNSG Kofi Annan. As quoted above, his 2000 Millennium Report challenged the international community to respond to humanitarian crises. The doctrine then had a gestation period within the international community (but outside the UN), lead initially by an independent expert commission. The commission's role is considered below. The gestation period was carefully managed by Annan and his successor (Ban Ki Moon). The process is summarised by Annan: "I couldn't have done it inside. It would have been very divisive. And the member states were very uncomfortable because, as an organization, sovereignty is our bedrock and bible - here is someone coming with ideas which are almost challenging it. So I had to sow the seed and let them digest it but take the study outside and then bring in the results for them to look at it." (Weiss et al. 2005, p. 378). It was not until 2005 and 2009 that states were asked (within UN processes) to support R2P. This was achieved by consensus with states first adopting the 2005 World Summit Outcome Document which recognised and outlined R2P (A/60/L.1 2005). In 2009 Moon released a report on implementation (UNSG 2009). Following a General Assembly debate, states passed a consensus resolution reaffirming the 2005 commitment and taking note of the UNSG's report (UNGA Res 63/308).

In contrast, the GPE was the exclusive initiative of French lawyers and legal scholars. Once drafted with the assistance of international legal experts but with a small final drafting team, it received political support from French President Macron who announced an intention to "push the Pact through the United Nations General Assembly." (Kotzé and French 2018, p. 812). France subsequently hosted a launch summit in September 2017, which enabled France and supporting states, to promote what became the UNGA Res 72/277 mandating the OEWG and requiring the gap report. Analysis of OWEG consultations suggests the GPE's setback was (in part) the price paid by external experts not understanding the UN system. Some saw it as an exclusive initiative that attempted to 'skip the line'. Furthermore, they did not spend sufficient time communicating objectives to states and the global public, nor did they build the necessary alliances (ENB 2019c, p. 11).

4.3 Inclusiveness vs an Elite Eurocentric View

Following Annan's challenge to the international community, the Canadian government established a fully independent commission (the International Commission on Intervention and State Sovereignty—ICISS), under Australian and Algerian co-chairmanship. ICISS membership was geographically, ethnically, politically, professionally and ideologically diverse.[4] Members included former heads of state, cabinet ministers, long-serving UN officials, military generals, scholars and journalists. Thakur notes; "it is important also to include representatives from the worlds of practitioners and ideas... Ideas are the key drivers of history and a panel that does not include them will have a short life." (Thakur 2019, p. 51). Critically, members had different ideological starting points, with some pro and others anti-intervention. Nevertheless, they were able produce a unanimous report that was not merely a collection of clichés and platitudes. (Thakur 2011, p. 33). The ICISS's final report was both comprehensive, innovative and well researched (ICISS 2001). However, the extensive outreach programme that attended its work was of critical importance.

[4] Members were from: USA, Russia, Germany, Philippines, South Africa, Canada, Switzerland, India and Guatemala.

The ICISS held over 200 meetings and round tables in almost all continents and major capitals, engaging all sectors and seeking a cross-section of views (Thakur 2011). These consultations enabled it to listen to concerns and fears, 'road test' ideas and recommendations as they emerged, and enhance legitimacy for the final report (Thakur 2019). Engagement with developing states was particularly important given their newly attained sovereignty and negative experiences of colonialism, including 'not very benign interventions' (Evans 2015). In short, the ICISS had fulfilled its mandate to promote comprehensive debate on the issues, and foster global political consensus on how to move from polemics towards action, particularly through the UN (ICISS 2001, p. 81).

In contrast, the GPE was not the product of an inclusive process but rather represented an elite view very closely associated with France, and later EU member states. Proponents highlight the inclusiveness of the drafting process and the support of 135 jurists to a 'Jurists' Call for a Global Pact for the Environment', representing 40 nations. However, closer inspection reveals that 65% were from Europe and the USA, 9% from Africa, 12% from Asia and 13% from South America. The majority of signatories were from France (23) and the USA (15). A major report supporting the GPE was in the form of an expert legal document (Aguila and Viñuales 2019b). This reality, together with French political ambition as a champion of multi-lateralism (President Macron) and the Paris Agreement (Fabius), raise legitimacy issues. These factors together with the vote on UNGA Res 72/277 (discussed above), criticism over initial lack of engagement with the G77 plus China and a sense of some states that they were not being listened to, played into fears, especially those from the Global South. These fears included a loss of sovereignty, additional unfair burdens and double standards (ENB 2019c, p. 11). In these circumstances, it is hardly surprising that many states were unwilling to support a legally binding instrument.

4.4 *The Precedents*

R2P was directly connected to a contemporary understanding of the role of the State and the argument that: "[l]iving up to the responsibilities of sovereignty becomes in effect the best guarantee of sovereignty." (Deng 2010, p. 364). As noted above, it was also embedded in a modern reading of the UN Charter, together with the development of human rights within the UN system. However, R2P also employed constitutional precedents from the African and Indian continents and in this way developed legitimacy. R2P was "as rooted firmly in indigenous values and traditions than in abstract notions of sovereignty derived from European thought and practice. Many traditional Asian cultures stress the symbolic link between loyalty of citizens to sovereigns and duties owed by kings to subjects." (Thakur 2011, p. 211). Thakur also highlights the alignment between R2P and traditional African understanding of the duties of rulers towards subjects (Thakur 2011, p. 211).

The GPE includes preambular references to the 1972 Stockholm Declaration, 1982 World Charter for Nature, and 1992 Rio Declaration. Some text uses language from the Earth Charter (e.g.; ecological integrity and community of life). However, the primary reference document appears to be the Rio Declaration, which is consistent with the Pact's codification objective. What it does not do is address the critical normative issue of shifting understanding of state sovereignty, by drawing explicitly on relevant precedents such as the Earth Charter, IUCN Draft Covenant, a growing list of responsibility declarations and emerging trusteeship jurisprudence. To be fair, the proponents never claimed this level of normative ambition for the Pact. This is clear from the very conservative principles on the 'duty of care of the environment' (Article 2) and the requirement to 'cooperate in good faith' to implement the Pact (Article 18). The one exception is Article 1: "Every person has the right to live in an ecologically sound environment …". This does represent a normative development, but it is not explicitly linked to important legal precedents including national constitutions (Boyd 2012).

The failure to use precedents in support of normative development certainly deprived states and others of an important opportunity to rethink state sovereignty in an age of ecological crisis.

4.5 Norm Entrepreneurs, Brokers and Champions

The success for R2P is closely linked to the presence of norm entrepreneurs, brokers and champions. The first is a person/s who identifies a normative gap and then organises a process to fill this gap. Brokers find solutions and champions actively support the emerging norm and defend it. R2P had norm entrepreneurs from within the UN system including Kofi Annan, assisted by states people from Canada, Algeria, Australia, India and the Sudan. The ICISS performed the role of norm broker through its report and engagement process. UNSG Moon became a powerful norm champion, defending R2P and actively developing mechanisms for its implementation (Thakur 2019). He is considered pivotal to the success of the 2009 UNGA Resolution (Madokoro 2015).

In the case of the GPE, the norm entrepreneurs (*Le Club de Juristes* and Fabius) were from outside the UN system and primarily associated with only one nation, France. The French government performed the role of broker by fast-tracking or pushing the GPE into the UN system via the UNGA resolution mandating the working group process. France was also the norm champion as it held extensive bilateral consultations during the working group process. However, France was not prominent during the formal negotiations because it was represented by the EU (ENB 2019c). In short, the GPE did not have diverse, prominent and influential normative leadership.

5 Conclusion

The above discussion outlines selected factors considered central to the success of R2P and considers the extent to which these are present or absent in the instance of the GPE. This is a preliminary analysis, however the results indicate that the GPE and other innovative civil society initiatives can learn much from the history of R2P. More comprehensive research is warranted.

There is an urgent need to find transformative legal responses to the global ecological crisis, however we must progress in ways that don't risk the already tenuous legitimacy of states and the rule of international law. This loss of legitimacy is much closer than it might seem. UNSG Guterres recently visited New Zealand, where he met with student climate change protesters. He was questioned about enforceability of the 2015 Paris Agreement, replying that it relied upon naming and shaming. The students responded by laughing at him. In doing so, today's youth demonstrated derision for the UN system, the authority of states and for international law. As Thunberg has urged: if you can't find solutions within the system, change the system (Thunberg 2018). While we might agree that a revolution is needed, how it is achieved is another matter.

References[5]

Aguila Y, Viñuales JE (2019a) A global pact for the environment: conceptual foundations. Rev Eur Comp Int Environ Law 28:3–12
Aguila Y, Viñuales JE (eds) (2019b) A global pact for the environment: legal foundations. C-EENRG, Cambridge
Axworthy L, Rock A (2009) R2P: a new and unfinished agenda. Global Responsib Protect 1:54–69
Biniaz S (2019) The UNGA resolution of a "Global Pact for the Environment": a chance to put the horse before the cart. Rev Eur Comp Int Environ Law 28:33–39
Bosselmann K (2015) Earth governance: trusteeship of the global commons. Edward Elgar, Cheltenham
Boyd D (2012) The constitutional right to a healthy environment. Environ Sci Policy Sustain Dev 54(4):3–15
Brunnée J, Toope SJ (2008) Norms, institutions and UN reform: the responsibility to protect. J Int Law Int Relations 2:121–137
Deng FM (2010) From sovereignty as responsibility to the responsibility to protect. Global Responsib Protect 2:353–370
Evans G (2015) The evolution of the responsibility to protect: from concept and principle to actionable norm. In: Thakur R et al (eds) Theorising the responsibility to protect. Cambridge University Press, Cambridge
Fabius L (2017) Letter from President of the French Constitutional Council to legal experts (24 March, 2017)
Fernández EF, Malwé C (2019) The emergence of the 'planetary boundaries' concept in international environmental law: a proposal for a framework convention. Rev Eur Comp Int Environ Law 28:48–56

[5] All websites accessed on 24-11-2019.

Fishel S (2018) Is climate change a threat multiplier? R2P and environmental disasters. https://www.e-ir.info/2018/04/24/is-climate-change-a-threat-multiplier-r2p-and-environmental-disasters/
French D, Kotzé LJ (2019) 'Towards a Global Pact for the Environment': international environmental law's factual, technical and (unmentionable) normative gaps. Rev Eur Comp Int Environ Law 28:25–32
Global Centre for the Responsibility to Protect. http://www.globalr2p.org/about_r2p
GPE Newsletter – 'Setback for the Pact', 27 May 2019
GPE.org: https://globalpactenvironment.org/en/the-pact/where-are-we-now/
Kotzé LJ, French D (2018) A critique of the global pact for the environment: a stillborne initiative or the foundation for Lex Anthropocenae? Int Environ Agreement Law Econ 18:811–838
Kotzé LJ, Kim RE (2019) Earth system law: the juridical dimensions of earth system governance. Earth System Gov 1:1–12
Madokoro D (2015) How the United Nations secretary-general promotes international norms. Global Responsib Protect 7:32–55
Slaughter AM (2005) Security, solidarity, and sovereignty: the grand themes of UN reform. Am J Int Law 99(3):619–631
Thakur R (2011) The People vs The State. United Nations University Press, Tokyo
Thakur R (2019) Reviewing the responsibility to protect: origins, implementation and controversies. Routledge, London
Thunberg (2018) https://eco-age.com/news/greta-thunbergs-best-quotes
Tigre MA (2019) World leaders debate on the future of the global pact for the environment. http://ccsi.columbia.edu/files/2019/09/Global-Pact-Event-Tigre.pdf
UN.org. https://www.unenvironment.org/people/laurent-fabius
Voigt C (2019) How a 'Global Pact for the Environment' could add value to international environmental law. Rev Eur Comp Int Environ Law 28:13–24
Weiss TG, Kuele (2011) Whither R2P? https://www.e-ir.info/2011/08/31/whither-r2p/
Weiss TG et al (2005) UN voices: the struggle for development and social justice. Indiana University Press, Bloomington

Reports, Documents and Resolutions

Common Home for Humanity (2019) Why do we need an Earth System approach to Guide the Global Pact for the Environment. https://wedocs.unep.org/handle/20.500.11822/27702
Earth Trusteeship Appeal (2019) https://earthtrusteeshipplatform.org/appeal/
ENB (2019a) Earth negotiations bulletin (21 January 2019), 35:1 (for all reports: https://enb.iisd.org/enb/vol35/)
ENB (2019b) Earth negotiations bulletin (22 March 2019), 35:2
ENB (2019c) Earth negotiations bulletin (25 May 2019), 35:3
Global Pact for the Environment (2017) Le Club de Juristes: https://globalpactenvironment.org/uploads/EN.pdf
ICISS (2001) Report of international Commission on intervention and state sovereignty, the responsibility to protect
Millennium Report of the Secretary General (2000) We the peoples: the role of the United Nations in the 21st century, https://www.un.org/en/events/pastevents/pdfs/We_The_Peoples.pdf
Oslo Manifesto for Ecological Law and Governance (2016) https://www.elga.world/oslo-manifesto/
Report of the ad hoc open-ended working group established pursuant to General Assembly Resolution 72/277 (May 2019)
UNGA Resolution (2009) A/RES/63/308
UNGA Resolution (2018) A/RES/72/277
UNSG (2009) Implementing responsibility to protect http://www.globalr2p.org/resources/283
UNSG (2018) 'Gaps in International Environmental Law and Environment-related Instruments: Towards a Global Pact for the Environment' UN Doc A/73/419

The Role of Trusteeship in Earth Governance

Klaus Bosselmann

Abstract The call for institutionalizing Earth trusteeship cannot easily be reconciled with state sovereignty. The concept of state sovereignty emerged at a time of great distances and absolute national autonomy. However, in a globalized, interconnected world utterly depending on the integrity of Earth's ecological systems, absolute territorial sovereignty is counterproductive and potentially life threatening. The chapter argues that the time is right for reconceptualizing state sovereignty. Sovereignty includes not just fiduciary and trusteeship obligations towards the state's own citizens, but also towards humanity at large and Earth as a whole. The UN Agenda 2030 with its Sustainable Development Goals offer a window opportunity for institutionalizing Earth trusteeship at international and national levels. A critical tool for achieving this has been the adoption of the "Hague Principles for a Universal Declaration on Responsibilities for Human Rights and Earth Trusteeship" in the Peace Palace, The Hague, on the day of the 70th anniversary of the Universal Declaration for Human Rights (10 December 2018).

1 Introduction

The 1987 Brundtland Report is known for having introduced the concept of sustainable development. Less known is this sentiment expressed at the beginning of the report (Brundtland Report 1987, p. 1):

> In the middle of the 20th century, we saw our planet from space for the first time. Historians may eventually find that this view had a greater impact on thought than did the Copernican revolution of the 16th century (…). We can see and study the Earth as an organism whose health depends on the health of all its parts. We have the power to reconcile human affairs with natural laws and to thrive in the process.

K. Bosselmann (✉)
Faculty of Law, University of Auckland, Auckland, New Zealand
e-mail: k.bosselmann@auckland.ac.nz

L. Westra et al. (eds.), *Ecological Integrity in Science and Law*,
https://doi.org/10.1007/978-3-030-46259-8_20

Reconciling human affairs with natural laws? Inspired by a vision of Earth? What does sustainable development have to do with Earth? Seemingly nothing if we look at references to sustainable development in multiple international agreements, policy statements or the 2015 Sustainable Development Goals (SDGs). They all describe "human affairs" such as well-being, equity and prosperity in almost complete isolation from their non-human, natural context. Even climate change is seen as a threat purely to human affairs, never mind the integrity of the atmosphere as an ecological system (Bosselmann 2019, ch. 5). The narrative of humans as measure and benchmark for anything that matters on Earth is still the same as 30 years ago.

Or is it? Arguably, the narrative is changing, not dramatically, but noticeably just as the Brundtland report predicted: studying Earth as "an organism whose health depends on the health of all its parts" has become a major subject in natural and social sciences. Earth system science assumes a holistic view of the dynamic interaction between the Earth's spheres and their many constituent subsystems (Kump et al. 2011; Ehlers et al. 2006; Colin Prentice et al. 2012; Lenton 2016). It provides us with a new narrative in which humans are in the middle, but not in the centre of ecological interactions. So are we finally waking up to the reality of one Earth system of which human and non-human systems are part and parcel?

Political and legal philosophies have increasingly been influenced by Earth-centred thinking. The drafting of the Earth (!) Charter during the 1990s is one indication of this, the ever-growing body of literature on Earth jurisprudence, Earth law and Earth governance another. Today, hardly any environmental lawyer would dismiss the need for a global, Earth-centred approach to the design of law and governance. This is despite or perhaps because of the omnipresence of "sustainable development". Ultimately, the Earth system will determine the fate of human systems (including legal systems), not the other way round.

The system of international environmental law has been organised and implemented by nation states. In itself this wouldn't be a problem, were it not for the fact that nation states have largely functioned as self-referential systems with little regard to commonalities outside their national boundaries. Earth is taken for granted, a legal nullity (*res nullius*) that has no voice in national or international decision-making. The world's legal systems are simply not geared for listening to Earth (who of course speaks to us in her own voice).

Yet, many people have become aware of this kind of deafness. They want to give Earth voice and expression in the form of stewardship and guardianship. This morality underpins the concept of Earth trusteeship.

Trusteeship is a form of governance that requires a person or entity to act (as 'trustee') on behalf and for the benefit of another person or entity ('beneficiary'). Arguably, the very institutions that act on behalf of and for citizens, i.e. states and governments, have trusteeship responsibilities. Protecting the rights of human beings ('human rights') and citizens ('citizen rights') requires the state to act as a trustee of its citizens (Benvenisti 2013). It couldn't work any other way. Similarly, protecting the integrity of Earth's ecological systems requires the state to act as a trustee for the Earth (Sand 2013).

Earth trusteeship is a term combining the morality of guardianship with the strong origins and rhetoric of enforcement associated with trusteeship as a legal concept. By transferring moral discourses around guardianship into strong and binding international law, there is an opportunity to overcome the current disconnects between problematic state practices and systematic change.

This chapter aims to show that Earth trusteeship, while undisputedly a challenge to unfettered state sovereignty, does not act entirely in opposition to it, and in fact can draw on its foundations for strong actions in favour of ecological sustainability. However, the current structures that surround state sovereignty are inadequate for facing ecological challenges, therefore need to be reframed and implemented in an eco-centric, rather than anthropocentric way.

Firstly, we need to distinguish state sovereignty from other international institutions and describe how it is affected in states' practice. It will then identify the shortcomings of these arrangements, before introducing earth trusteeship as a means to rectify environmental tensions in both domestic and international law. This can be forwarded by assurances that the current systems, such as those supported by the United Nations, do not entirely exclude trusteeship concepts, and in fact are structured in such a way that they may be beneficial to their implementation.

2 Institutional Barriers of Environmental Trusteeship

2.1 State Sovereignty

The current paradigm of sovereign nation states was established by the Treaty of Westphalia in 1648. Subsequently, the United Nations framework and its founding documents have upheld the concept as part of the modern international system. Article two of the UN Charter prescribes state sovereignty as the primary means by which the organisation operates.

The main reference point for this theory is that states are representatives of their citizens and governments are elected by those citizens, instigating a form of social contract (Horn 2016). Nationality becomes the main identity of individuals within states, and of individual states within the international system (Bosselmann 2015, p. 72), so generally speaking states exclusively pursue their own interests (Bosselmann 2016, p. 24). As a result, international law, including international environmental law, operates through a system of multinational environmental agreements, subject to each state's sovereign decision to follow them.

From this, the system can be described as individualistic, fragmented, economy-focused, and utilitarian by its very nature (Bosselmann 2016, pp. 29–31), and falls short of any cohesive governance of the global commons or Earth as a whole. At UN level, there have been attempts to rectify this by promoting guiding doctrines, such as the 'common heritage of mankind', but these have remained limited by states strongly upholding their property rights and as such are generally passive, if not tokenistic. Additionally, the short-term political cycles through which states' democratic processes operate further inhibit any long-term commitments to environmental governance. Theories link this sovereign right of states and their political institutions to exploit their own resources directly to property rights (Barnes 2006, p. 34) even going as far as calling state sovereignty the 'international mirror' of private property rights (Bosselmann 2016, p. 134). Doubts about the integrity of democratic processes themselves also arise with these concerns (Bosselmann 2018, p. 3).

However, the essence of democratic processes fundamentally allude to a trust relationship. This claim is succinctly pursued by Eyal Benvenisti, who sees sovereignty as an exercise of personal and collective self-determination (Benvenisti 2013, p. 301), sovereign states as agents of humanity as a whole (Benvenisti 2013, p. 305), and sovereignty as currently entailing the power to exclude portions of global resources (Benvenisti 2013, p. 308).

As such, the concept of environmental trusteeship is not actually in conflict with institutions of state sovereignty. Instead, it is the ordering of the state over and above other interests that is problematic. But in order to somewhat equalise the shortcomings of a state-centred approach, a compromise is required which initiates less sovereignty so that states can be bound by their trusteeship obligations within a commons-centred approach, but more sovereignty as could evolve through environmental trusteeship (Bosselmann 2016, p. 143).

2.2 Design of International Environmental Law

These days, international environmental law is framed by calls for urgent responses to a changing climate, impending mass species extinction, and many other symptoms that can be collectively described as an Earth system crisis. These responses should reframe states as a means to effectively govern the neglected global commons, rather than an end in themselves as a sovereign entity (Bosselmann 2015, p. 2). Currently, legal governance institutions and their mechanisms are distanced from the inherent moral purposes that they should serve, resulting in anthropocentric policy and frameworks at both the domestic and international levels (Bosselmann 2015, p. 34). The challenge here is that environmental law holds a central non-human component which causes the paradox that it operates with boundaries that ecosystems by their very nature ignore.

Some weaknesses of the state-centred system are self-imposed in that while states continue to insist on the primacy of their sovereignty and perpetuate an anthropocentric, individualistic, and capitalist system, they are losing their power to regulate (Bosselmann 2015, p. 9). Economic interdependence and globalisation has meant that instead of states, transnational corporations have the majority share of resources which enables their manipulation of the neoliberal global economy. This means that those wealth-pursuing agendas are forwarded, to the detriment of the global commons and on a wider scale, to the detriment of the environment. In fact, western liberal democracies are still struggling to even acknowledge ecological challenges (Bosselmann 2015, pp. 13–22). This is exemplified by the mere existence of debate about the human-induced nature of global warming. This translates to a reductionist approach to problem-solving, which is often reactive or attempts to find basic connections between specific causes and effects (Bosselmann 2016, p. 85).

The policy that emerges is directed to the short term and is simplistic and immature as a result, failing to see the forest for the trees, or appreciate the level of complexity that has arisen by fault of a staunch utilitarian and economic perspective on the environment. A paradox that emerges here is that this western neoliberal system has evolved under democratic checks and balances, so in theory should reflect the sentiment of the public. However, it is evident that those being the most affected are the least heard and coupled with unequal distributions of wealth this evokes environmental justice issues at the wider international level that this concept applies to 'developed' and 'developing' states, and domestically to disadvantaged communities. The current version of democracy needs to be revisited so as to promote policy moving from the bottom. This would mean that individuals, as activists and advocates, become decision-makers, driving notions of environmental trusteeship as an exercise of sovereignty.

Contemporary international environmental law holds two central pillars—the notion of sustainable development, and the precautionary approach. The latter is only implemented weakly against the standards of the former, so that the recognition of the rights of non-human forms and the inclusion of future generations remain only promises. Nature has been commodified to a critical point, reminiscent of a 'tragedy of the commons' situation. Policy approaches favouring increased productivity are particularly harmful given that technological advancement has allowed for the exploitation of previously uncontemplated domains. Private property rights represent an attempt to regulate the overuse of such domains, but these are incompatible with the behaviour of diverse ecosystems interacting within a finite planet. Actions occur at a local level within a state's own territory and may appear to have limited negative effects, but they may create reactions and tipping points in other locations and for the macro ecological system known as the commons. Not only is such a human-nature divide fatal to the biosphere, but it also erodes human relationships with each other (Bollier and Helfrich 2019). These complexities are difficult for current environmental law to govern effectively given the problematic economic focus and false assumption that limitless growth is possible. Application of the precautionary principle occurs in competition with individual nations' usage rights without any legal consequences, and weak enforcement mechanisms mean this is allowed (Benvenisti 2013, p. 316).

The second central pillar of contemporary international environmental law is 'sustainable development', as introduced and defined in aforementioned Brundtland Report. The term recognizes concern for the poor and future generations alongside the environment, however it fails to acknowledge any limit to growth based on a finite natural environment. The focus of the concept is anthropocentric, evoking a weak approach to sustainability. Further to this, the SDGs under the 2030 Agenda lack cohesion and do not refer to planetary boundaries. While the SDGs are somewhat reflective of the first cosmopolitan reaches of environmental law, they have also been critiqued as a lost opportunity where a paradigm shift to an ecocentric model of commons governance could have occurred (Montini and Volpe 2015, p. 142).

Another downfall of the operation of the legal international environmental system is its obsession with procedure. This, in addition to the fragmented and individualist characteristics of that system, means that urgent action is difficult to implement, and longer-term solutions become diluted. Because only nation-states can voluntarily enforce the law against other states, the system is also overwhelmingly political which overshadows actual environmental goals and tends to exclusively advance human interests. Essentially this means that instead of humans evolving within the Earth system, we are evolving the system ourselves (Bosselmann 2016, p. 26).

It is therefore established that, given planetary limits, the current system of international environmental law is inadequate. It focusses on economic and social systems rather than ecological systems and has resulted in the continuous deterioration of the Earth system as a whole. Such failure can be summarized as, *"a mismatch between global systems and the different interests of diverse states. In such a context, the sum of national interests is not always sufficient to meet systemic requirements"* (Bosselmann 2016, p. 144). The different interests of diverse states are upheld over the systemic requirements of the global system, maintaining a weak approach to sustainability based on the lowest common denominator. In contrast, a 'strong' approach would model action around Earth system needs, resulting in a 'safe operating space' for humanity (Raworth 2012; Magalhaes et al. 2016).

However, state-centrism is unlikely to be dissolved in the near future. Instead, it needs to be connected to a real concern for future generations and legal recognition of the global commons including Earth. This new, more nuanced concept would include the state as a trustee, imposing inalienable normative duties under a rationale that combines state sovereignty with the preservation of the integrity of Earth's ecological systems. This would include powers to resolve disputes through an enforcement branch independent of the United Nations and as such incurring wider and stronger accountability without relying on the assent of states.

3 The States as an Environmental Trustee

State sovereignty is too robust a pillar in international law to be disrupted in the near future. Consequently, states' governance mandates should be expanded so that their actions appropriately reflect global and intergenerational concerns of their citizens based on ecological values (Bollier 2008). In this way, the state is the only body available to act as trustee in what is already primarily a trust relationship (Benvenisti 2013). Benvenisti premises this under the idea that state sovereignty is an exercise of personal and collective self-determination and that as such states contract to act as agents of humanity as a whole (Benvenisti 2013, pp. 301–305). There are also numerous examples where states, in their current position as sovereigns, have accepted the social contract behind this responsibility for citizens and non-citizens alike, as well as the environment. What remains is for states to pragmatically recognise their ethic of care, as could be instituted under a reconciled version of Earth Trusteeship.

This would mean that the state would be taken to accept, as part of its democratically-enforced social contract, further fiduciary duties for the global commons and future generations. Under this arrangement, members of society, as among the beneficiaries of this trust relationship, could charge for damage to common property, and by extension, be rewarded for protecting it. These direct legal consequences constitute the necessarily strong environmental action that has real potential to rectify the current crisis. Criddle and Fox-Decent premise that such an arrangement of social policy would subordinate economic and social policy to ecological interests, imposing a standard of ecological integrity in the form of a bottom line (Criddle and Fox-Decent 2009, p. 331). If this were to involve rights-based legislation, for example, economic and social rights would only be advanced where they did not unreasonably interfere with the right to a healthy environment. Likewise, notions of property rights would be transformed so instead of existing in an individual capacity, the human community's common heritage of life is prioritized. This model instigates the planetary boundaries framework for limiting state sovereignty, which again steers global governance away from its current anthropocentric values.

However, such a model cannot be effectively implemented if it is to exist as a mere framework under an international legal structure. The democratic nature of civil societies requires citizens to accept their agency as guardians and trustees, so that systematic changes are underpinned by a new environmental ethic coming from the bottom up (Bosselmann 2016, p. 129). Trusteeship principles should be encouraged through all experiences with the environment so that they are contemplated in the same fundamental way that principles of fairness arise in all social lives. This involves attracting human consciousness, though the new youth movement (*Fridays for Future, Extinction Rebellion*) suggests that such an attitude is very much in motion, and potentially only lacks acknowledgement from higher-level authorities. Hopefully, such acknowledgement could forward a system of non-competitive cooperation under a cosmopolitan, holistic, eco-centric worldview.

4 Towards Earth Trusteeship

To a degree, Earth trusteeship already fits international environmental law. The United Nations advocates a fiduciary-based approach (Bosselmann 2015, pp. 198–232) and can offer cohesion, at least of identity, because it has already begun instituting formal mechanisms of accountability. In terms of jurisdiction, the political motivation of most states is also foreseeable (Bosselmann 2018, p. 3), so that while currently a range of obligations fall largely on those states, it is likely that they will be willing to instigate wider accountability. It is therefore the most appropriate institution to strengthen and expand so as to develop trusteeship governance. There is a question, though, as to the extent of legal recognition of the environment that would need to take place before such commitment could be conceded by states. Options to overcome this are wide-ranging, including *grundnorms* of ecological integrity, rights of nature and future generations, and legal personhood of individual ecosystems. These will be discussed further below.

Currently missing from mainstream discourse around international environmental law is an appropriate theory of the state. This can be forwarded through more scholarship to demonstrate that the state inherently has a trusteeship function. An important part of such a theory is the recognition of two types of land ownership. i.e. property and the commons, or *res communes* under Roman law. This latter type ensures permanent use of all through effective mechanisms of protection. Historically, the commons have been vulnerable to overuse (Hardin 1968), but this was due to the lack of effective trusteeship governance. Without it, property rights will take over and the commons will eventually suffer. This phenomenon was greatly accelerated in the age of capitalism where property rights have virtually replaced any trusteeship governance of the commons (Bosselmann 2015, p. 107; Bosselmann 2016, pp. 129–136).

By returning to the various political theories that describe how the state gains its legitimacy, trusteeship obligations become apparent and indeed a key function of the twenty-first century state. Essentially, the democratic state comes into being through election by its people, meaning that in order to retain its authority it must uphold the fiduciary obligations through which those same people can hold it accountable. These can also be derived from social contract theory, in particular if associated with the public trust doctrine of North America and other notions of trusteeship, such as guardianship or, in the case of New Zealand, the tikanga Māori concept of *kaitiakitanga*. The assumption that ultimately humanity aims to survive can be safely made (Bosselmann 2016, p. 53) requiring states to adopt a position as agents of humanity as a whole (Benvenisti 2013).

This theory of the state therefore concludes that redefining the concept of state sovereignty to include trusteeship obligations is possible due to democracy. Its normative basis is the *grundnorm* of ecological integrity in the same vein as the existing *grundnorm* of human rights. Both not only legitimize the state as the law enforcing authority, they also necessitate it as a guarantor and trustee. Without trusteeship obligations to protect human beings and the integrity of ecological systems, the state lacks any legitimacy.

On this basis, we can turn to the means by which Earth trusteeship may be institutionalized in international environmental law. It should be noted that trusteeship governance is inherent in documents such as the Earth Charter which can be considered an ethical framework for Earth governance. However, to remain in that nascent status it needs to be defined and translated into (constitutional) law. Drawing from the democratic theory of the state, the starting point for this is recognizing the active interests of citizens and making value-based decisions from these. A variety of options have been proposed or attempted, from a weak 'greening' of decision-making, through to systemic transformation. What has been established in the literature is that nations are causing the degradation of the commons by their own actions and common ecosystem services are provided by resources within national territories. As a result, an individualistic system based on usage rights is ineffective for solving ecological problems. We need to draw on cosmopolitanism for an approach that overcomes these anthropocentric boundaries and represents people instead of states.

It is proposed that already established concepts around 'common concerns' and 'common heritage' are appropriate starting points for developing Earth trusteeship governance. The discourse surrounding the Common Heritage of Mankind doctrine focusses on intergenerational equity and governance of the commons to bring this about. A broader framing of a 'common heritage of life' is important here as it takes an ecocentric perspective for the protection of the global commons and the Earth system (Taylor 1998, p. 298).

Another option could be to expand existing frameworks of normative rights by introducing, for example, rights of nature. In New Zealand, this has already played out in the introduction of legal personhood status to natural beings, such as the Whanganui River (Whanganui River Claims Settlement 2017). They can be seen as potentially groundbreaking depending on the actual institutionalisation and practice of *katiakitanga* (trusteeship) governance.

New Zealand's legal arrangements would fit easily into an international legal system that was based on wider environmental mandates. It could be implemented by co-management arrangements under states' multilateral environmental agreements and enforced by an international environment court. Existing structures are not inherently opposed to this framing. The United Nations Security Council could, for example, be enhanced to include an environmental framework for legitimacy, and by employing an authority such as a global environmental ombudsman or a Commissioner for Future Generations liaising between the various states and institutions could be harmonious.

The entire premise of such arrangements is that states, with their current expectations of full and unfettered sovereignty, are unable to see and legislate further than their anthropocentric values. Their responsibilities need to be reframed, which is not entirely contrary to a retention of simplistic state sovereignty. Framing such sovereignty against trusteeship obligations, the state would have full legal responsibility and a fiduciary duty to act in good faith in order to affect the best interests of the beneficiaries. The beneficiaries in such a scenario include all people, future generations and the Earth system.

Within the UN system, the now defunct UN Trusteeship Council could be reconceived into an environmental body to institute member states as trustees for the global commons (Stoddart 2011, p. 37). In this way it would act for administrative and executive purposes. Part of this role would be to reduce the political tensions that arise inevitably under the current paradigm of state sovereignty (Biermann 2011, p. 5).

There is in fact an established tradition of UN institutions with a trusteeship mandate including the UN Trusteeship Council, the World Health Organization (WHO) with respect to public health and—somewhat ironically—the World Trade

Organization (WTO) with respect to free trade. A number of other UN or UN-related institutions with weaker trusteeship functions exist also. Quite obviously, states have been capable of, expressively or implicitly, creating international trusteeship institutions. These developments—and in particular the existence of supranational organizations such as the European Union—demonstrate that sovereignty of states can be transferred to international levels.

The UN Trusteeship Council could quite easily be revived as an Environmental Trusteeship Council following proposals by the Global Governance Commission in 1995 which were supported by a number of states and particularly championed by the former UN Secretary General, the late Kofi Annan. A combination of environmental activism and new political alliances (e.g. between particularly motivated progressive states) can make a crucial difference. Chances are that such combined effort will be very powerful as our global ecological, financial, political and democratic systems continue to disintegrate.

We cannot expect trusteeship governance being initiated by the "top", i.e. the UN and its member states themselves, but rather by forces outside the system, in particular global civil society. To this end, we can build on many years of activism and proposals for institutional change. Nor should we advocate states to be in charge of running and controlling global trusteeship institutions such as a World Environment Organization or a Global Atmospheric Trust or a UN Earth Trusteeship Council. Rather their governance must be jointly formed by representatives from global civil society, UN and states with an equal say in decision-making.

Essentially, there are three requirements of the institutionalization of earth trusteeship in a form that does not entirely overthrow state sovereignty. Firstly, the nation state must recognize a legal responsibility for the global commons and Earth as a whole. Secondly, redesign these so that rather than belonging to states or individuals, they belong to all of humankind. This would be sufficient to shift the current anthropocentric focus of environmental law to an earth-centred approach. Thirdly, this law must identify and hold to account the inalienable normative duties of states as reconciled with intra- and inter-generational equities. Whatever form this took, it would inevitably refocus state sovereignty towards challenges to environmental priorities. However, this is necessary in order to encourage consistency and trust in all interactions with the environment and to begin attempts to restore ecological integrity (Wood 2007, 2013; Sand 2004; Sand 2013, p. 21; Barnes 2001, 2006).

The inextricable connections between rights, responsibilities and trusteeship are at the heart of the Earth Trusteeship Initiative, a consortium of social and environmental justice organisations aiming for legal innovation. Its first objective has been achieved now, i.e. a joint declaration on trusteeship responsibilities for both, human rights and sustainability called the "Hague Principles for a Universal Declaration on Responsibilities for Human Rights and Earth Trusteeship" (Earth Trusteeship Initiative 2018). The "Hague Principles" were launched at the Earth Trusteeship Forum in the Peace Palace, The Hague, on 10 December 2018 to mark the 70th anniversary of the adoption of the Universal Declaration of Human Rights.

The Hague Principles define three different kinds of responsibilities, i.e. those that humans have for each other in order to protect human rights, human responsibilities for Earth and state responsibilities for protecting human rights and the Earth. All these responsibilities have their moral base in our belonging as human beings to the community of life. Their legal expression is trusteeship. Humans and states act as trustees and can in this way be held accountable for any failures in meeting their responsibilities.

The objective now is to garner world-wide support for the idea that humanity must accept trusteeship responsibilities for the community of life.

5 Conclusion

Earth governance breaks with the traditional rule that care for the environment ends at national boundaries. Protection of the integrity of Earth's ecological systems must be a domestic concern as much as an international concern. In fact, not states, but the well-being of Earth must determine degree and quality of environmental protection. Such an Earth-centred viewpoint forces states into the logic of Earth trusteeship.

At the 2019 Global Economic Forum in Davos and again at the United Nations in New York, the Prime Minister of New Zealand, Jacinda Ardern, made the point that to be on the right side of history world leaders must embrace guardianship of the Earth. How true! Only then it may be possible to turn things around and avoid global ecological collapse.

References

Barnes P (2001) Capitalism 2.0: who owns the sky? Our common assets and the future of capitalism. Island Press, Washington

Barnes P (2006) Capitalism 3.0: a guide to reclaiming the commons. Berret-Koehler, San Francisco, p 34

Benvenisti E (2013) Sovereigns as trustees of humanity: on the accountability of states to foreign stakeholders. Am J Int Law 107:295

Biermann F (2011) Reforming global environmental governance: the case for a United Nation's Environment Organisation (UNEO). Stakeholder Forum, Earth System Governance Project, and VU University Amsterdam, p 5

Bollier D (2008) The commons: a neglected sector of wealth creation. In: Heinrich S (ed) Genes, bytes and emissions: to whom does the world belong? Henrich Boll Stiftung, Berlin

Bollier D, Helfrich S (2019) Free, fair and alive: the insurgence power of the commons. New Society Publishers, British Columbia

Bosselmann K (2015) Earth governance: trusteeship of the global commons. Edward Elgar Publishing, Cheltenham, p 72

Bosselmann K (2016) The planetary integrity project: creating a safe operating space through law and governance. New Zealand Center for Environmental Law, Auckland, p 24

Bosselmann K (2018) Reclaiming the global commons: towards earth trusteeship. In: Martin B, Te Aho L, Humphries M (eds) Responsability: law and governance for living well with the earth. Routledge, London, p 3

Bosselmann K (2019) The atmosphere as a global commons. In: Jaria-Manzano J, Borràs S (eds) Research handbook on global climate constitutionalism. Edward Elgar Publishing, Cheltenham. chapter 5

Brundtland Report (1987) Report of the World Commission on environment and development: our common future. https://sustainabledevelopment.un.org/content/documents/5987our-common-future.pdf. Accessed 24 Nov 2019

Colin Prentice I, House J, Cornell S (2012) Understanding the earth system: global change for application. Cambridge University Press, Cambridge

Criddle E, Fox-Decent E (2009) A fiduciary theory of Jus Cogens. Yale J Int Law 34:331

Earth Trusteeship Initiative (2018) The Hague principles for a universal declaration on responsibilities for human rights and earth trusteeship. http://www.earthtrusteeship.world/the-hague-principles-for-a-universal-declaration-on-human-responsibilities-and-earth-trusteeship/. Accessed 24 Nov 2019

Ehlers E, Krafft T et al (eds) (2006) Earth system science in the anthropocene: emerging issues and problems. Springer, Berlin

Hardin G (1968) The tragedy of the commons. Science 162:3859. 1243–1248

Horn C (2016) Kant's political philosophy as a theory of non-ideal normativity. Kant-Studien 107(1):89–110, pp 89–110

Kump L, Kasting J, Crane R (2011) The Earth system, 3rd edn. Pearson Education, Upper Saddle River

Lenton T (2016) Earth system science: a very short introduction. Oxford University Press, Oxford

Magalhaes P, Steffen W, Bosselmann K et al (2016) The safe operating space treaty: a new approach to managing. Our use of the earth system. Cambridge Scholars Publishing, Newcastle

Montini M, Volpe F (2015) Sustainable development goals: much ado about nothing. Environ Liability 4:141–147, p 142

Raworth K (2012) A safe and just space for humanity; can we live within the doughnut? Oxfam discussion paper. Oxfam International, Oxford

Sand P (2004) Sovereignty bounded: public trusteeship for common pool resources. Global Environ Politics 4(1):47–71

Sand P (2013) The rise of public trusteeship in international environmental law. Global trust working paper series 03/2013. http://globaltrust.tau.ac.il/publications/. Accessed 24 Nov 2019

Stoddart H (ed) (2011) A pocket guide to sustainable development governance, 1st edn. Stakeholder Forum, Commonwealth Secretariat, London, p 37

Taylor P (1998) An ecological approach to international law: responding to the challenges of climate change. Routledge, London, p 298

Wood MC (2007) Nature's trust: a legal, political and moral frame for global warming. Environ Aff Law Rev 34:577

Wood MC (2013) Nature's trust: environmental law for a new ecological age. North Carolina University Press, Durham

Legislative Citation

Te Awa Tupua (Whanganui River Claims Settlement) Act 2017

Environmental Damages and Armed Conflict

Gabriella Colao

Abstract The concerns about the environment and the associated damages to natural resources in times of armed conflict have emerged in distinct phases of the history reflecting the existing connection between the progress achieved through technology and the application of such scientific knowledge with the way they have transformed the concept of military operations. In the last century important protecting rules and criteria have been implemented to protect civilians, to safeguard those who no longer take part in hostilities, to distinguish civilian objects from military ones, and also to protect the environment as a whole in armed conflicts. However the issue of damages in times of war still remains highly controversial. The absence of an integrated and holistic approach in the normative framework applicable to armed conflict makes the environment extremely vulnerable and at risk, regardless of the incredible expansion of environmental law since the 1960. The three subsystems of international law, namely humanitarian law, international criminal law and international environmental law, although have incredibly broadened the legal basis for the protection in wartimes, present conflicting criteria, especially in determining liabilities for damages in armed conflict. Nevertheless balancing military operational requirements with environmental protection and security should not be viewed as an impossible task, even in conflicts. There is a critical need to increase the protecting efforts of nature and ecosystems both in peacetime and in war, exploring concepts that nature itself has developed and looking at new legal instruments from the perspective of common good and global commons, a revisited notion of national interest and of responsible sovereignty.

The concerns about the environment and associated damages to natural resources in times of armed conflict have emerged in distinct phases of the history reflecting the connection between the progress and application of scientific knowledge with the

G. Colao (✉)
International Organization, Naples, Italy

L. Westra et al. (eds.), *Ecological Integrity in Science and Law*,
https://doi.org/10.1007/978-3-030-46259-8_21

way they have transformed the concept of military operations. At the same time environmental factors were driving changes in the political relations among States.

Already in the late 1800 existing treaties were regulating the issue of dangerous substances, chemicals, or gas in war. However, the major concerns were directed to the protection of people, and only indirectly addressing the problem of contamination of the environment. The Hague Conventions of 1899 and 1907 were the first multilateral treaties regulating the conduct of warfare and were largely based on the Lieber Code, signed and issued by US President Abraham Lincoln to the Union Forces of the United States on 24 April 1863, during the American Civil War. The 1907 Conventions consisted of thirteen treaties and one declaration. The majority of them though had already been violated during World War I. Poisonous gas was in fact forbidden under the IV Convention, but had been largely used.

Additionally, important protection rules, complementing the Hague conventions and seeking to protect people who are not or are no longer taking part in hostilities, as the sick and wounded of armed forces on the field, wounded, sick, and shipwrecked members of armed forces at sea, prisoners of war, and civilians were universally agreed in the aftermath of the World War II under the Geneva conventions. Following the tragedy of the World War II and as a successor of the League of Nations that proved to be totally ineffective, the United Nations were established precisely with the intention to guarantee a durable peace through friendly relations among States, through cooperation and through the decision to ban war for ever. Nevertheless war is an inevitable component of the political relations. As Von Clausewitz would say "*war is merely an extension of diplomacy by other means*". Competition for resources and power continue to forge in many ways international relations. To preserve peace and security, among the goals of the United Nations the development and implementation of International Law is a fundamental component for the expansion of the rule of law. International environmental law should be part of an effective contribution to this goal not only for the specific scope of preservation and protection of the environment. It would also better support an enhanced level of cooperation among nations. This chapter aims at discussing the effectiveness of the current framework of environmental law in the context of armed conflicts.

During the 1970 a new phase of international action associated with increased risks concerning the environment began. Pollution, loss of marine resources, of wetlands, nuclear testing and preventive conservation in general started to be more and more at the centre of public discussions with the aim of demanding appropriate international rules to protect natural resources. Several more regional and global treaties were negotiated, broadening this interest and the promise of the international community to bear a much more responsible attitude in limiting or preventing the dangers to the environment. The debate on the use of certain weapons in those years affecting the environment quite strongly animated the public opinion.

In 1972 the General Assembly of the United Nations endorsed the Stockholm Declaration ("Report of the United Nations Conference on the Human Environment" 1972), whose principle 1 affirms that "*man bears a solemn responsibility to protect and improve the environment for present and future generations*", reminding us of our intergenerational responsibilities. Moreover, principle 21 of the said declaration strongly announces that States have "*the responsibility to ensure that activities within their jurisdiction or control do not cause damage to the environment of other States or of areas beyond the limits of national jurisdiction.*" Principle 26 of the Stockholm declaration places the accent on the importance for States of recognizing their role and responsibility in advancing in the level of cooperation. Because of the specific concerns regarding the impact of armed conflict and of certain dangerous weapons on the environment, it reminds the international community that "*man and his environment must be spared from the effects of nuclear weapons and all other means of mass destruction. States must strive to reach prompt agreement, in the relevant international organs, on the elimination and complete destruction of such weapons*".

These major concerns about environmental security and the need for a more holistic approach in environmental issues led the international community to enter a third phase that prompted several more agreements, among which the 1992 Framework Convention on Climate Change, the 1992 Convention on the Conservation of Biological Diversity and in particular to the 1992 Rio Declaration ("Rio Declaration on Environment and Development" 1992). The latter document echoed the utmost importance of the prevention aspect. In principle 24 it is in fact said that "*warfare is inherently destructive of sustainable development. States shall therefore respect international law providing protection for the environment in times of armed conflict and cooperate in its further development, as necessary*".

Since the atomic bombings of Hiroshima and Nagasaki on 6 August 1945, nuclear weapons have been detonated over two thousand times for testing and demonstrations. They produce enormous blast energy creating a devastating shock wave causing a huge amount of heat and lethal ionizing radiations, radioactive debris and dust propelled by winds high into the atmosphere, and then settling to earth as radioactive fallout. In regards to human life it is such an enormous toll in destruction, death, injury, and sickness of incomparable scale with any other single weapon with a disproportionate effect under the rules established by applicable conventions, and certainly with long-lasting effects on the environment, such as large quantities of contaminated wastes on land, or sea and in the air and contaminated plants and animal life. The repercussion on the environment were legally evaluated in 1996 when the ICJ stated in its Advisory Opinion on the Legality of the Threat or Use of Nuclear Weapons that "*the environment is not an abstraction but represents the living space, the quality of life and the very health of human beings, including generations unborn*" ("Advisory Opinion on Legality of the Threat or Use of Nuclear Weapon" 1996). Though the ICJ observed in their opinion that the principal purpose of environmental treaties and norms is the protection of the environment in times of peace, in remarking that "*the environment is under daily threat and that the use of nuclear weapons could constitute a catastrophe for the environment*", it reminded our consciences that even in the most tragic human event like war there are limits bypassing which the possibilities for conservation of a healthy environment may be lost, the sustainability of life and the survival of ecosystems could be degraded for ever.

Restriction in the use of force and in the means and methods of warfare is the scope of the branches of international law regulating and limiting it through principles and thresholds, whose violations may entail international criminal liability. As stated in other decisions the Court held, "*in the field of environmental protection, vigilance and prevention are required on account of the often irreversible character of damage to the environment and of the limitations inherent in the very mechanism of reparation of this type of damage*" (International Court of Justice 1997). This vision is reflected in the 1982 World Charter of Nature, adopted by United Nations General Assembly on 28 October 1982, urging us to realize that "*nature shall be secured against degradation caused by warfare or other hostile activities*" (World Charter of Nature). Such degradation may determine the fall into an irreversible cycle where man cannot control it anymore.

As a result of this global awareness of the impact of man's activities on earth, the incredible expansion of environmental legal instruments during the last 60/70 years has not given rise to a coherent system of protection. Regardless of the efforts of part of the scientific community in showing us the other face, the dark side of the industrialization, of technocracy and of the capitalistic paradigm based on the assumption of unlimited growth and on the absolute right to the extraction of natural resources, and in parallel of the effects of these concurrent factors in war, environmental law has not found its way to make the environment at the core of the protection issues and to articulate consistently in its body the concerns of scientists. Although there is no agreement in the scientific community, a new term, "Anthropocene", is informally and widely used to refer to a proposed geological epoch in which the significant human impact is profoundly altering earth's geology and ecosystems, including but not limited to anthropogenic climate changes. Applied science in particular has always been used in support of new means of warfare, with the aim of enhancing security and projecting power. However the more science advances, the more relevant becomes the protection dilemma. Food war, bacteriological weapons, contamination, techniques of modification of the environment as a weapon, all these are a constant threat. The major concerns derive by the use of any of these means by organizations or failed states.

To respond to these dilemmas international humanitarian law, known also as law of armed conflict, and international criminal law in the context of war, both in their relationship with international environmental law have broadly expanded. The origin of the Law of War dates to centuries ago. Sun Tzu's famous book, *The Art of War*, was written in the fifth century before Christ, but many examples of legislations can be traced in ancient times, already showing the constant struggle of humankind to control war. In wars as in peace laws of humanity must continue to exist and must not be forgotten. Environment must also not be forgotten.

A more schematic legal approach started with the Hague stream which generated, as already commented, several Hague conventions from 1899 to 1907, dictating laws and customs of war on land, prohibiting launching of projectiles and explosives from balloons, rules on neutrality, limits to naval warfare. Its main goal was to limit means and methods of warfare, though preservation of the environment was not the primary goal, but only the effect of an indirect protection. War and military operations in preparation of a war do not happen in a separate universe or in a totally separated world, but are conducted in a specific space, the environment. Such space, as rightly specified by the ICJ, is not an abstraction but a fundamental element of life. However the protection was very limited and compliance with the rules was and still is a difficult objective, particularly in consideration of the fact that the environment is merely seen as an object.

Since then through the waves in the 1970 we have acquired a deeper awareness that the protection and preservation of the natural environment is fundamental to the survival of the human kind. Direct considerations, and not merely indirect one, must be regarded in means and methods of warfare under different assumption of the notion of environment. The development of international law is quite slow and modern methods of warfare do not easily fall in old rules. Moreover it is also rather unreasonable to conceive the environment in isolation from other parts. Biology and many branches of science have advanced our perception and understanding of the environment as an interconnected whole. This is not properly reflected in the law which continues to regulate segments of the environment, segments of life. Lacking an understanding of the implications of an incorrect interpretation of the notion of environment from a legal standpoint requires precaution. There is in fact never scientific certainty as to the effects on the environment of certain military operations and therefore all feasible precaution must be considered and taken to avoid incidental damages to the environment (Rule 44).

The second important more schematic approach is the one arising from the 1859 Battle of Solferino in Italy, a major battle in world history, following which the determination of Henry Dunant, founder of the Red Cross, gave birth to the conference of nations and to the setting out of the process that led to the four Geneva Conventions in 1949. The Conventions were updating the antecedent treaties of the 1929, and were followed by the adoption of two Additional Protocols in 1977 and a third one in 2005. The treaties were establishing the standards of international law for the humane treatment in armed conflict, of international and non-international nature, such as rules for the protection of the wounded, shipwrecked, prisoners of war, or in general those who do not take part in hostilities, protection rules for civilians and civilian objects. Both streams, the Hague and the Geneva streams, are now technically under the same body of law, known as Law of Armed Conflict or International Humanitarian Law. It continues to develop in order to respond to questions that modern warfare poses, while the basic principles remain a constant, namely Military Necessity, Distinction, Humanity, Proportionality, and Precaution.

Other legal instruments complement the protection of the environment in warfare, such as the 1977 Convention on the Prohibition of Military or Any Other Hostile Use of Environmental Modification Techniques (ENMOD Convention), which prohibits parties from engaging in military or any other hostile use of environmental modification techniques, having widespread, long-lasting or sever effects as means of destruction, damage or injury to any other party.

Additionally, in the wave of expansion of the general body of international environmental law, there is an extensive list of new international treaties that would be impossible to cite, of which some examples are the 1982 United Nations Convention on the Law of the Sea (UNCLOS) Montego Bay, the 1980 Convention on Certain on Prohibitions or Restrictions of Certain Conventional Weapons which May Be

Deemed to Be Excessively Injurious or to Have Indiscriminate Effects also known as 1980 Inhuman Weapons Convention), the 1993 Chemical Weapons Convention, the 1959 Antarctic Treaty and the 1967 Outer Space Treaty, the ICRC multi-volume explanation of customary IHL that includes 161 rules, three of which are dedicated to the protection of environment and natural resources; last but not least the Draft Convention on the Prohibition of Hostile Military Activities in Protected Areas, strongly advocated by the International Union for the Conservation of Nature (IUCN). The last document was developed right after the Gulf War in 1990–1991 in response to the risks of environmental and ecosystem damages in armed conflict.

Together with all these legal texts, soft law is somehow a significant guidance in interpreting this complex picture, providing indications or suggestions in many aspects of environmental protection or enforcement means, through Resolutions of the General Assembly, ICRC Guidelines for Military Manuals and Instructions on the Protection of the Environment in Times of Armed Conflict, statements and regional documents and so on. It is worth adding that several relevant tribunals or arbitration tribunals and courts have also been established, giving rise to extensive jurisprudence in matters of environmental law as in matters of international criminal responsibility.

The progressive expansion in scope and in the protection under international law has created a complicated system, where separations undoubtedly are real, due to the all the existing subsystems of international law. The result is a deep fragmentation of international law where the real problem is not which body of law is applicable to a particular context, but how a specific problem can be solved by inferencing rules from different sources of law. This is in fact the central problem for damages in armed conflict, explicitly the applicability of different rules deriving from different sources of international law, sometimes with conflicting or unclear criteria, including those pertaining to international environmental law, to military activities and warfare. Julian Wyatt has elaborated this vision and affirms that the problem of environmental damage lies at the intersection of the relevant branches of the law, specifically international humanitarian law, international environmental law and international criminal law (Wyatt 2010).

The relevance of the overlapping of applicable bodies of law is visible in some situations. While in times of conflict Law of Armed Conflict is considered the applicable body of law, to some extensions it suspends other bodies of international law. Nonetheless to mitigate these effects some non-nuclear weapons States have affirmed that environmental agreements should continue to apply. Rules deriving from the principles present in the Stockholm and Rio Declarations should not be considered inapplicable as they both precisely refer to environmental risks in war. Somehow by-passing the argument, some nuclear States affirm that "*without addressing the general question of the applicability of multilateral environmental agreements during conflict, [...] such agreements (Principle 21/Principle 2) could not be construed as prohibiting the threat or use of nuclear weapons because they did not address the nuclear weapons per se and could not be construed as containing an implied prohibition on their use*". However what is necessary and proportionate in conformity with the principles of Law of Armed Conflict in terms of human protection, must similarly be considered in environmental deliberations. Therefore, it is quite obvious that in situations where the subsystems of international law seem to be overlapping, if not contradicting themselves, the environment remain undoubtedly at risk of poor or irrelevant protection.

The other problematic aspect is that it is similarly rather obvious that Law of Armed Conflict has been highly anthropocentric for a long time and that the environment has been recurrently regarded only as a civilian object, instrumental to human life. The issue of wartime environmental damages and the criminalization of serious violations has indeed introduced in many cases personal criminal liability for military commanders. Nonetheless in the context of the Additional Protocol I (1977) relating to the Protection of Victims of International Armed Conflicts, art. 35 (3) and art. 55(1) are the only provisions directly speaking of environmental damages. Art. 35 refers to the means and methods of warfare, prohibiting the use of which may cause widespread, long-term, and severe damage to the natural environment. In this article it was the first time a direct protection of the natural environment per se, and not just under the general definition of civilian objects, was made. It applies to intentional damages and collateral ones as well. On the other hands, art. 55 includes a prohibition of the use of methods or means of warfare which are intended or may be expected to cause such damage to the natural environment and thereby to prejudice the health or survival of the population. And in this case it does only refer to environment as a civilian object.

In comparison the Rome Statute establishing the International Criminal Court (ICC) on July 2002 envisions twenty-six violations where the Court has jurisdiction in the event they are perpetrated in armed conflict. Not only they are considered grave breaches of the law in respect to the limited framework set by the Additional Protocol I, but the view of the doctrine is that by far this criminalization may contribute to a major shift in the compliance with the law applicable in war. However, the evaluation of the environmental damages may be considered by far the most difficult component to establish as it requires two distinct mental elements, the intent to launch an attack and the actual knowledge of the consequences of "widespread, long-term and severe" damages to the environment in comparison with the expression used in the Additional Protocol I "*may be expected to cause*". The element of knowledge is linked to the cumulative consequences in order to establish the violation, which is a quite tough threshold.

The document which clearly addresses environment is the Environmental Modification Convention (ENMOD), formally the Convention on the Prohibition of Military or Any Other Hostile Use of Environmental Modification Techniques, approved by the General Assembly of the United Nations in its resolution 31/72 of 10 December 1976 in the same period of the 1977 Additional Protocols to the Geneva Conventions. In application of paragraph 2 of the said resolution, the Secretary General decided to open the Convention for signature and ratification by States at Geneva, Switzerland. The Convention was signed by 48 states; as of January 2018, the Convention has 78 state parties. Art.1 provides the scope, which is the prohibition of environmental modification techniques for changing the environment through a deliberate manipulation of natural process, namely "*the dynamics, composition or structure of the Earth, including its biota, lithosphere, hydrosphere and atmosphere, or of outer space*". Undoubtedly, not every modification techniques of the environment accounts for a prohibited conduct, but those that are military and hostile in nature, whether for offensive or defensive purposes, intentional and causing destruction, damage or injury to another State. The term hostile implies the use of the environment as a weapon causing certain effects, in primis environmental damages.

There is a list of prohibited activities at art. 2, such as "*earthquakes, tsunami, an upset in the ecological balance in the region, changes in weather patterns (clouds, precipitation, cyclones of various types and tornadic storms), changes in climate patterns, changes in ocean currents, changes in the state of the ozone layers, changes in the state of ozonosphere*". The threshold for damages in the case of the ENMOD Convention is narrower although it uses the same adjectives, "widespread", "long-term" and "severe". The big difference lies in the fact that while the Additional Protocol I lists them cumulatively, the ENMOD lists them alternatively. The other relevant difference is about the intentionality in the "expected" damage which is greater in the case of the Additional Protocol I, also comprising purely unintentional and incidental damages (Dinstein 2001).

However, one of the important questions remains the threshold of the three elements, widespread, long-term and/or severe, in order to assess the violation of international law as the legal texts fail to define these terms.

To add to this complexity, what is unclear is that the term natural environment does not have a commonly accepted definition. The definition of environment reflects different approaches in doctrine. While some provide a narrow definition that strictly refers to natural resources and in terms it covers ecosystems and biological diversity, others tend to lean towards a broader definition, which embraces also aesthetic values, such as cultural heritage, landscape and environmental amenity. The ENMOD, without a notion of environment, gives only a list of prohibited activities.

However, while the environment in general terms is considered a "civilian" object as opposed to the military in nature, it is recognized that it should not be subject to attacks, unless it is used as a military object by a party to the conflict. Under this dual use and/or the fact that it is used to hide its military nature, it may lose the protection offered under Law of Armed Conflict to civilian objects. In such case the military necessity is the driving principle and no damage should be procured that does not provide a clear military advantage. If there is indeed an assessment of such nature, the proportionality principle is a requirement.

Despite of all the challenges arising out of this complex picture of international law and its subsystems, balancing military operational requirements with environmental protection and security are not impossible tasks. It is normal practice to include environmental protection in training to military forces. In many military manuals environmental protection is covered. Moreover, in the planning process of military operations the presence of subject matter experts, such as military engineers or environmental experts, are required. The development of national or multinational doctrines in environmental protection has been quite impressive in recent years. Clearly the compliance with international law to avoid damages, liabilities and claims is not an easy task during an armed conflict or even in the transition from conflict to peace as demonstrated by many examples of case studies on post conflict reparations. The environmental protection in post conflict times constitutes a tremendous challenge for the management and disposal of assets. Countries that have gone through a devastating war and are in transition to peace, especially unstable peace, somehow don't list environmental diligence as a priority.

We cannot conclude this short paper on environmental damages in armed conflicts without touching the issue of processing environmental claims arising out of such events. According to the International Law Commission, there is an internationally wrongful act of a State when: (a) conduct consisting of an act or omission is attributable to the State under international law; and (b) that conduct constitutes a breach of an international obligation of the State (International Law Commission (I.L.C.)). However, the responsibility of the State in cases of violations of the law of armed conflict goes beyond the normal principles of State responsibility. Article 3 of Hague Convention IV states: *A belligerent party which violates the provisions of the said Regulations shall, if the case demands, be liable to pay compensation.* It shall be responsible for all acts committed by members of its armed forces.

The State shall be responsible for acts or omissions committed by its organs in violation of provisions of *jus in bello,* the legal framework described for the legitimate use of the force, that have caused environmental damages. In the example of Iraq's invasion of Kuwait in 1990 the Security Council condemned the act as an illegal invasion and violation of the *jus ad bellum* rule, prohibited by the UN Charter. The conflict caused extensive damages to the environment and around the Gulf waters and originated a huge amount of environmental claims. Environmental damage of three types occurred: 1—damage to the marine environment as large quantities in the order of ten million barrels of oil were released by Iraq from the Sea Island Terminal in Kuwait affecting 1500 Kms of coasts; 2—damage to an extensive area of the environment, including the air, caused by the burning by Iraqi forces of over 500 oil wells in Kuwait; and 3—land degradation arising out of the aerial bombardment, existing minefields, the construction of defensive fortifications such as trenches, and the land campaign conducted by armed forces, both the Iraqi and the coalition forces.

The Security Council Resolution 674 (1990) considered Iraq liable under international law it for any loss, damage or injury arising in regard to Kuwait and third States, and their nationals and corporations, as a result of the invasion and illegal occupation of Kuwait by Iraq and invited States to collect information about potential claims. Resolution 687 (1991), that was adopted after the end of the fighting, again reiterated Iraqi's responsibility descending by the illegal invasion and occupation of Kuwait as well as the connected liability under international law for any direct loss, damage, including environmental damage and the depletion of natural resources, or injury to foreign governments, nationals and corporations. Following this decision, it was also established a Compensations Commission which would settle all the claims under a fund established with Iraqi's revenue of oil sales. It must be reminded though that the commission is not a court, but it acts as a subsidiary organ of the Security Council. Its primary scope was to conduct the extensive work resulting from the submission, processing and settlement of an impressive amount of more than of more than 2.6 million of claims (Greenwood 1996).

Environmental security is more and more at the core of global actions because of the daily direct and indirect threats to the environment. Consequently more attention focuses on how the applicability of environmental law should consider a coherent approach in military operations too and how damages are to be effectively treated in armed conflicts. In view of the current climate changes threats to peace and security may develop and competition for scarce resources will multiply the risks of conflicts and the associated risks of environmental damages; it will also be the major factor of displacement and migrations in the future. The repercussions of climate change will be perceived at multiple levels, from global to national, regional and local levels.

Advancement in the current view of environmental protection has been interpreted as a move towards a more appropriate notion of ecological integrity. The current legal framework does not suffice for its fragmentation and old view. "*A new conception of life has in fact emerged at the forefront of life, a unifying view that life's biological, cognitive, social and ecological dimension*" (Capra 2015). The profound change of metaphors is emerging, from seeing the world as a machine with single, spare parts that environmental law should fix, to understanding it as a network, a system and therefore in terms of connectedness, relationships, patterns and context. The concept of connectivity recognizes that habitats and species function best as a part of large, interconnected network that is maintained and protected for nature by including people. While connectivity conservation is a strategy and a new area for law (Worboys et al. 2010) in the specific context of preservation and connection of protected areas, this methodology could be used as a model in other areas of law to increase the applicability of a systemic approach.

There is no possibility within the current legal and fragmented system, which is still linked to a mechanistic and very sectoral paradigm, to move from environmental law to an approach of ecological integrity, which frankly would be the only possible realistic option of survival. Among the ramifications of international law applicable to armed conflict the issue of environmental damages is also still difficult to address given the current picture. Of course, this is just the top of the iceberg. The existing regulatory framework should be fully re-evaluated and a new description should be attempted moving from the formal compliance of rules applicable through their specific subsystems to a new scenario where regulations must be read through an approach similar to the one offered by the connectivity conservation law.

While security is at the core of military operations, in recent years there has been also a shift in the concept of security, from a state centric notion of defence to a broader concept entailing human security up to an even broader idea under development where universal responsibility is recognized in terms of protection, although linked to specific cases. However universality plays a role. The whole purpose of universal jurisdiction, for example, is rooted in the concept that some international crimes go beyond the jurisdiction of a State and can be prosecuted by any State. Time has come to elevate environmental law to another dimension that includes in this discussion a more holistic and universal view also in regards to the environment security as well. In his encyclical letter (Enciclycal Letter Laudato Si of the Holy Father Francis on Care for Our Common Home) Pope Francis affirms that we need a more consistent and responsible global reaction. If the human environment and the natural environment deteriorate together, there is no way we can win this battle with the available tools, and more and more conflicts will arise which in turn will deteriorate the environment even more.

One primary condition is to move from a position of pure national interests that the majority of states pursue to a position where it is accepted the notion of "common good", a political and philosophycal category, of the human kind. In light of the acceptance of a discussion around this principle then, as we may recognize that everything, including the earth, the oceans, the climate, the water should be driven by the principle of common good, the second condition would be to find ways to legally determine the category of "global commons", not to be confused with common good, and how to implement efficient systems of governance of global commons, namely resources that lie outside the political reach of any one nation State. The third condition could be to start at least discussing around the concept of earth rights, though there are many difficulties in constructing a legal subjectivity. In moving from a legal dimension to the dimension of philosophy of law the notion of common good, coming from the Tomistic tradition of natural law, confirms its essential characteristic, which is indivisibility. This unique constituent is capable of influencing law and politics and allows the translation of an appropriate theory of sovereignty. Without this element of an indivisible common good, politics and law will always struggle in finding solutions to the constant conflictuality, to grave violations, and the use of force will remain the remedy used in international relations. To this aim the legal notion of states as environmental trustees has been proposed leading to a real paradigm shift. So far sovereignty sets limits to environmental protection, but ultimately it should be environmental protection to set the limits to state sovereignty (Bosselmann 2015).

References

"*Advisory Opinion on Legality of the Threat or Use of Nuclear Weapon*", in ICJ Reports Year 1996, p. 226 e seq. (241, para. 28 and 29)

Bosselmann K (2015) Earth governance: trusteeship of the global commons. Edward Elgar, Cheltenham

Capra F (2015) Laudato Sì – the ecological ethics and systemic thought of Pope Francis, 22 June 2015. http://www.fritjofcapra.net/laudato-si-the-ecological-ethics-and-systemic-thought-of-pope-francis/. Accessed 15 Oct 2019

Dinstein Y (2001) Protection of the environment in international armed conflict. Max Planck UNYB 5, p. 541

Enciclycal Letter Laudato Si of the Holy Father Francis on Care for Our Common Home. Available at http://www.vatican.va/content/francesco/en/encyclicals/documents/papa-francesco_20150524_enciclica-laudato-si.html. Accessed 3 Oct 2019

ENMOD Convention available at https://treaties.un.org/. Accessed 3 Nov 2019

Greenwood C (1996) State responsibility and civil liability for environmental damage caused by military operations. In: Grunawalt RJ, King JE, McClain RS (eds) Protection of the environment during armed conflict, International Law Studies vol 9. US Naval War College, Newport https://digital-commons.usnwc.edu/ils/vol69/iss1/43/
International Court of Justice, Judgment on Gabčíkovo-Nagymaros Project (HungarylSlovakia), 25 September 1997. https://www.icj-cij.org/files/case-related/92/092-19970925-JUD-01-00-EN.pd. Accessed 15 Oct 2019, [140]
International Law Commission (I.L.C.), Draft Articles on State Responsibility, Part I, Art. 1 at https://legal.un.org/ilc/texts/instruments/english/draft_articles/9_6_2001.pdf
"Report of the United Nations Conference on the Human Environment" Stockholm Declaration (Stockholm 516 June 1972) UN Doc A/CONF.48/14/Rev.1, part II, Principle 21
"Rio Declaration on Environment and Development" UN Conference on Environment and Development (Rio de Janeiro 3-14 June 1992) UN Doc A/CONF.151/26, vol. I, Principle 2
Rule 44. Due Regard for the Natural Environment in Military Operations
Worboys GL, Francis WL, Lockwood M (2010) Connectivity conservation management: a global guide. Earthscan, London, ecopy available at
World Charter of Nature. Available at https://www.refworld.org/docid/3b00f22a10.html. Accessed 2 Oct 2019
Wyatt J (2010) Law-making at the intersection of international environmental, humanitarian and criminal law: the issue of damage to the environment in international armed conflict. Int Rev Red Cross 92(879):593–646

The manufacturer's authorised representative in the EU is Springer Nature Customer Service Centre GmbH, Europaplatz 3, 69115 Heidelberg, Germany. If you have any concerns regarding our products, please contact ProductSafety@springernature.com

Printed and bound by CPI Group (UK) Ltd, Croydon, CR0 4YY
30/09/2026
02225689-0010